D1446850

# ••• THE •••
# TYRANNOSAURUS PRESCRIPTION

# ISAAC ASIMOV

# THE TYRANNOSAURUS PRESCRIPTION

## AND 100 OTHER ESSAYS

PROMETHEUS BOOKS
BUFFALO, NEW YORK

 Inquiries should be addressed to Prometheus Books, 700 East Amherst Street, Buffalo, New York 14215, 1-800-421-0351; in New York State, 716-837-2475.

Library of Congress Cataloging-in-Publication Data

Asimov, Isaac, 1920-
The tyrannosaurus prescription and 100 other essays / Isaac Asimov.
p. cm.
Bibliography: p.
ISBN 0-87975-540-7
I. Title.
PS3551.S5T97 1989
814'.54—dc20

89-8486
CIP

To my dear wife, Janet,
who makes every day a happy day.

# Contents

## SCIQUEST



## "FOREWORD BY ISAAC ASIMOV"

# Introduction

It is my job, and has been for fifty years, to put my ideas and opinions on paper, and to get them published for the world (or for as much of it as may be interested) to read.

During the first three years of my efforts, I sent the products of my typewriter to various possible outlets in the faint hope that someone might be willing to offer them a home. More often than not, they came back to me. But some sold, and in 1941 I published a story called "Nightfall," which made a hit. With that I must have found my voice, for the rejections stopped.

In the first fourteen years of my writing career I published science fiction only. I have never stopped writing science fiction and continue to write it, quite successfully, to this day, but in the early 1950s, I began to write nonfiction as well, first on science, and then on virtually *anything.*

My reputation grew as a writer who could turn out reliable material very quickly and on almost any subject. Consequently, I never had to peddle anything after "Nightfall." I wrote only on order. And eventually that began to keep me perpetually busy. I have now published well over four hundred books of all kinds (including science fiction, of course) and over three thousand shorter pieces of one sort or another.

Most people who commission essays from me are often content with relatively short ones, perhaps because they suspect that readers are impatient these days with so much else to occupy their minds—from sports, to television, to computer games. As a result, I have found that I have accumulated 101 pieces with an average length of about 1200 words (some shorter, of course, and some longer).

So it occurred to me that I would put them together as a kind of collection into which people could dip almost at random when they have some minutes to spare. Since the pleasant people at Prometheus Books always seem prepared to welcome a collection of miscellaneous essays from me (this is my third with them, after *The Roving Mind* and *Past, Present, and Future*), I sent it to them under the whimsical title, *The Tyrannosaurus Prescription,* and here it is.

Incidentally, I have tried to select the pieces in such a way as to avoid "thought duplication," but this is not always possible. I can't come up with a brand-new point of view for each essay; so, if I speak on two subjects that are somewhat allied, I am liable to repeat some of my points occasionally. Please forgive me when I do.

I have also tried to impose some sort of order on a collection of essays that, after all, I wrote without ever once worrying whether they could be strung together in some sensible way. Naturally, then, my imposed order is imperfect, but I have managed to put them into seven categories. If on reflection you decide that you could have made a better and more logical arrangement, you are probably right.

But I've tried.

# Part I
# The Future

---

As far as my nonfiction output is concerned, I am best known as a "futurist," so a good proportion of the requests I get are for some insight into one or another aspect of the future.

I don't pretend that my view of the future is necessarily correct. In fact, if the misadventures of past futurists are any guide, my vision will prove laughably incorrect. However, I am stuck with it, and, who knows, I may not be too far off the mark.

Incidentally, people who want me to write on the future usually want me to write on those subjects that interests *them.* I am never successful at convincing them that I don't know enough about a particular subject to write about it. In the first place, they refuse to believe me—they probably suspect that I'm just playing hard to get in order to raise my fee . . . which I never do, honest! And, in the second place, how can I plead ignorance forcefully enough when I hate to puncture my own reputation as someone who "knows everything"? After all, I make my living out of that misconception.

As a result, I let myself be talked into doing essays on the future of handicrafts or the future of chemical engineering despite my pious wail that I don't even know anything about the *present* of chemical engineering. (Most frightening of all was contemplating the future of marriage.)

Any of you, by the way, are free to criticize anything I say. If you do, I may learn something.

The essay on the future of chemical engineering is, by the way, the longest one in the book: 5000 words long. I might not have included it were I not so pleased that I was able to write on the subject at all. (The people I wrote it for were pleased, too.)

# 1
# Our Future in Education

How can we imagine what public education will be like in 2076, the time of our nation's tricentennial? Let's first try to imagine what *society* will be like.

Perhaps our civilization will have collapsed by then, under the weight of a harrowing population increase and fatal shortages of food and energy. There would be starvation and misery. Billions would die, and the remnant that would survive would be forced to exist in an environment badly, and perhaps permanently, damaged by civilization's death-throes. There would be *no* public education, except for what some could glean from hoarded books salvaged from the ruins of cities.

But let us instead suppose that civilization survives. What are the requirements of survival? First and foremost, we must learn to limit our numbers by some means other than a rise in the level of death and destruction. Humankind must lower its birth rate to that of the death rate or below.

If that is done, and if other, lesser problems are solved, the twenty-first century should see civilization moving ahead as science and technology continue to advance.

A low-birth-rate society, however, is one that will produce an enormous change from that to which humankind has always been accustomed. Combine a low birth rate with the further-extended life-span we can expect if science and medicine continue to advance, and it will be obvious that in the twenty-first century we will have a population with a smaller percentage of young people and a larger percentage of middle-aged and old than ever before in history. In fact, the twenty-first century will be the first era in history in which the old will outnumber the young.

This is a change we can see coming now. In the United States, the steadily increasing percentage of the old has now made those over sixty-five into a formidable voting force. What's more, we are becoming, increasingly, a nation that has its finances organized about the pensions, Medicaid, and social security that are enjoyed by so many and looked forward to by so many more.

As some have pointed out, there seem to be more and more unproductive oldsters being supported by the labors of a smaller and smaller reservoir of productive youngsters. This fact has been used by those who argue *against* lowering the birth rate. The supply of young must be kept up, the argument goes, or civilization will collapse under the weight of the old.

But if the supply of young is kept up, civilization will collapse anyway. What, then, is the solution? Might it not lie in education?

Traditionally, public education is confined to the young. Children understand this, and if there is any inconvenience to school, it is attributed by them to their crime of being young. They come to realize that one great reward of growing up is to become free of the prison of school. Their goal is not to be educated, but to get out, to be sprung.

Similarly, adults are sure to associate education with childhood, with something they have fortunately survived and escaped from. The freedom of adulthood would be sullied if they were to go back to educative life-habits associated with childhood. As a result, many adults sink into vegetating ignorance. They are as unembarrassed at having forgotten what little algebra and geography they once learned as they are at no longer wearing diapers.

In a society in which those over forty outnumber those under forty, this vegetating ignorance should not be allowed to continue. Education must no longer be confined to the young. The young must not look forward to its completion, nor should the old look back on it as thankfully over. For all people, education must be made to seem a requirement of human life as long as life endures. Mental and creative vigor should accompany the physical vigor that medical advance will allow. Human beings can then remain "productive," in our present understanding of the word, until advanced age.

But is this possible? Will the time come when people so enjoy being educated that they will be willing to engage in it, on and off, all their long life? Why not, if they can learn what interests them and not what someone in authority says they ought to learn, interested or not? It will mean that we must turn education from fixed curricula into the direction of personal taste.

After all, as time goes on (civilization surviving), the world will grow increasingly automated and computerized. The dull, repetitive work of the world, both physical and mental, will be done by mechanical devices, and to human beings will be left the task of creation. The world will increasingly become a world of leisure. Education will have to become leisure-oriented.

To an increasing extent, the world will be running itself, and the very notion of "productive" and "unproductive" human beings will wither. Naturally, then, people will be able to go their own way. There will always be those who will *want* to learn computer technology, or to engage in scientific research, or to devise new educational procedures. If anything, I should guess there would be more people voluntarily anxious to help supplement the mechanical running of the world than are needed.

And the rest? Some might be interested in writing, composing, painting, or sculpting; some in sports or travel; some in show-business of one sort or another; some in sleeping in a hammock, if they can bear the boredom of it.

It will be the task of education to help each individual to find within himself that activity that will make him most happy, his life most filled with interest—and that will then surely contribute to the happiness and interest of others as well.

In personalized education, one thing may well lead to another. A youngster who wishes to learn baseball and nothing else may become interested in reading in order to read about baseball. Or she may wish to learn arithmetic in order to calculate baseball statistics—and in the end find she likes mathematics more than baseball.

Indeed, may we not expect interests to change with age as a matter of routine? At age sixty, why might not someone suddenly decide to study Russian or take up chemistry, or venture into chess or archaeology or brick-laying? Why may not someone in later life turn from stamp-collecting to nuclear physics—or vice versa? And through all these choppings and changings, why should it not be a person's inherent right, always, to receive help from the public education system?

But how can we manage an education system that will be so individual, so one-to-one, as to allow each person to be educated according to his own bent and desire, whatever that may be?

Well, suppose that communications satellites become more numerous and far more versatile and sophisticated than they are today. Suppose that it is not microwaves but the far more capacious laser light that is used to carry messages from earth to satellite and back to earth.

Under these circumstances, there would be room for many millions of separate channels for voice and picture, and it can be easily imagined that every human being on earth could have a particular wavelength assigned to him, as now he might be assigned a particular telephone number.

We can then imagine that each person has a private outlet to which can be attached, when he wishes, a personal computerized teaching machine. It would be a far more versatile and interactive teaching machine than anything we can put together now, for computer technology will also have advanced in the interval.

We can reasonably hope that the teaching machine, programmed for some particular field of study, will nevertheless be sufficiently flexible and versatile to be capable of modifying its own program (that is, "learning") as a result of the input of the student. In other words, the student can ask questions that the machine can answer and answer questions in a way the machine can evaluate. As a result of what the machine thus gets back, it can adjust the speed and intensity of its course of instruction, and can even shift in whatever direction student interest directs.

Nor need we suppose that the teaching machine be a self-contained, finite object (like a television set, for example). We can reasonably suppose that the machine will have at its disposal any book, periodical, or document in the vast central, encoded, computerized global library, and that the machine can use this information to modify its program. What the machine has, the student has, either placed directly on a viewing screen or reproduced on paper for more leisurely study.

Naturally, we can suppose that no human being will necessarily be a mere passive receptacle for information. Any human being, once guided along his path of interest, whatever it may be, is very likely to make advances of his own, which can be fed back into the machine and through that into the global library, so that every student becomes a teacher as well.

By the tricentennial, then (assuming that civilization survives), humanity and machine could be developing a profound symbiosis. Humanity could have a more enriched life and understanding than it ever would have achieved through man's unaided brain. The computerized teaching machine will become the mental telescope through which glories greater than we can now imagine will be seen.

# 2

# Filling the Brain Gap

There are five billion people in the world today. If you eliminate the young children, the very old, the infirm, and a few of those rich by inheritance, the rest are, in one way or another, working in order to ensure that they live as comfortably as possible.

What makes it possible for a human being to do useful work? Muscles and brain.

Human muscles, while useful, as we must all admit, are by no means unique in the world of life. There are muscles that are stronger, by far. In the exertion of sheer power, a human being cannot compete with a horse, let alone with an elephant.

It is for that reason that human beings, having learned to domesticate animals, put the larger ones to work. Donkeys drew carts and oxen drew ploughs. And when the horse was finally domesticated, it turned out to be the most valuable working animal of all. (We still describe an assiduously industrious person as a "workhorse.")

In addition, human beings learned to bend the inanimate world to their uses, to invent tools that made their muscular efforts more effective.

You can carry water, after a fashion, in your cupped hands, but you can do so in larger quantities and more effectively in a pot. You can pry up a rock by sheer force, but you can do it more easily with a lever. You can drag a load across the field, but how effortlessly you can do it, in comparison, if you put it in a wheeled cart.

Beginning two hundred years ago, human beings found new sources of energy in the steam engine, the electric generator, the internal-combustion engine. As machinery has continued to multiply and grow more complex, the load of muscular labor has been lifted from the backs of humanity, particularly in those parts of the world that have undergone industrialization. Today, in a nation like the United States, so little-used are human muscles that we must invent uses for them and "work out" in order to try to keep them in reasonable order.

The other tool used by human beings in their work is the brain—and here, at least, we *do* have something unique. There is no other living creature, alive now or in the past, that uses its brain as effectively as we do. It is by use of the amazing human brain that human beings can invent ingenious devices of all sorts, compose symphonies, write books, conduct and coordinate the labors of others, lead armies, make decisions, and penetrate the secrets of nature.

No other form of life can help us here. We are on our own.

To be sure, we can use our brains to devise inanimate help. We can invent writing to record our experiences and pass them on to the next generation. We can devise number systems and calculating devices from the abacus to the computer. We can think up a scheme of alphabetical organization, write reference books, and do many other things that will make it easier for the human brain to do its work. In the end, though, these are merely subsidiary aids; the essence of the job is left to the brain.

If we compare the human brain to the human muscles, we have no trouble deciding that the former is the more important. If we allow human muscles to become flabby, we become flabby human beings—but we are still human beings. What's more, if we are not too old or too far gone, a program of exercise can get us back into shape.

If we allow the human brain to become flabby and to lose its function, however, we lose something that is vitally human. To become a *stupid* human being is to become less than a human being. And, in all likelihood, that loss of easy brain utilization is something we can never recover.

Naturally, there is a conflict between brain and muscle. To begin with, a life spent exclusively in hard muscular work, under conditions that seriously underuses the human brain, produces *boredom,* which is another reason why it is better to have animals or machines do the work. Animals are far less likely to grow bored, and machines cannot grow bored.

Boredom is a serious ailment, incidentally, that marks the underfunctioning of the thinking process. Such underfunctioning quickly aggravates the boredom to the point of mental atrophy. A brain that is never given the opportunity to think is very likely to lose the ability to do so. (Hence the stereotype in older times of the stupid, animal-like peasant. They weren't born that way; they were fashioned so by the totally unstimulating life they were forced to lead. Certainly, many brilliant creative thinkers of today are the descendants of such peasants; nothing was wrong with the gene pool.)

Consequently, the freeing of human muscles by modern machinery has given an important impetus to human creativity. As the percentage of people forced to spend their lives in unskilled muscular labor decreases, the percentage who can contribute something creative is bound to increase.

And yet . . . there is an odd gap here.

Even in the most industrially advanced societies there remain jobs, too complex to be carried out by animals or machinery, that nonetheless seriously underuse the brain. We know that this brain gap exists, but what we have done so far to close it is insufficient.

A job on an assembly line might not force you into a life of unremitting hard labor, but it does make it necessary for you to do work that is repetitious, stultifying, unrewarding—and, worst of all, that leaves you without the necessity of thinking. (Do you remember Charlie Chaplin in *Modern Times*?)

Much office work is similar. The necessity of filing, of fetching and carrying, of typing, of doing all sorts of labor that is too skilled for a machine but too unskilled for the full functioning of a human brain produces, first, a sense of unhappy drudgery, and second, an escape from that unhappiness by the development of an anesthetized and incapable brain. (Naturally, all this depends on the nature of the work. As work grows more responsible and more diversified, the brain is more occupied and suffers less.)

The net result is that, even in the most technologically advanced portions of the world, there is a large element of totally uncreative individuals who have become fit only to do the low-thought drudgery that they, in fact, do.

It is easy for those of us fortunate enough to engage in creative labor to dismiss the faceless underclass by saying, "Well, that's all they're fit to do." I'm afraid, though, that the truth is, "That's all they've been *made* fit to do."

But now, in our generation, there have come the computer and the mobile computerized machines we call "robots." (The two are not essentially different: Computer plus mobility equals robot.)

Computers and robots represent as important an advance as the use of animals for work, as the invention of machines, and as the discovery of new power sources. For the first time we have devices that can do jobs that till now were beyond machines. For the first time we have a chance to fill the brain gap, to make it unnecessary for human beings to engage in mental drudgery—just as earlier inventions had made it unnecessary for them to indulge in physical drudgery.

It's simple to see that anything that a robot can do is unfit for a human being to do. To but it another way, if a human being does work that a robot can do, that work will end by making a robot of the human being.

Once again, with the coming of robots, there will be a great surge in human creativity.

But is that too easy a jump? Is that an overestimation of human capacities by a hopelessly overoptimistic individual? Can we really expect human beings to be generally creative? Is not creativity a rare and precious phenomenon that occurs only occasionally?

It depends on what you mean by *creativity*. If you're speaking of supreme genius—that of a Mozart or a Shakespeare or a Newton—then, yes, it is exceedingly rare. But what about moderate creativity? What about musicians who aren't Mozarts but who add to our heritage of pleasurable songs and compositions? What about writers who turn out to be less than Shakespeare but who are amusing and instructive nevertheless? What about scientists who will never be Newtons and Einsteins but who make an occasional useful discovery anyway?

Is even this too much to ask for? As we look around the world, can we convince ourselves that the great mass of noncreative people are only made so and that, in a different world, they would not exist?

Actually, we have gone through this once before in history. There was a time, not many centuries ago, when literacy was rare. The ability to make marks that indicate words and to do it rapidly and neatly, together with the ability to look at those marks and interpret them quickly and unmistakably, was clearly something for only the most agile of brains. Reading and writing was the mark of a scholar and the peasantry could not be expected to be literate any more than they could be expected to fly. Even the aristocracy, which could occasionally be educated into a form of literacy, were rarely able to do it well.

With the coming of the Industrial Revolution, things changed. The unskilled farming of the time did not really require literacy; but, as the people shifted from the farms to the mills and were put to work at complicated machinery, literacy became essential. For that reason, the nineteenth century saw the development of free public schools for the education of the population. And, behold, while reading and writing remained difficult for many, literacy became far more widespread that would have been dreamed possible only a century before.

Nowadays it is a scandal that there are millions of Americans who are functionally illiterate. No one believes, however, that this is because literacy requires a rare kind of brain. The feeling is that our system of education is at fault.

In the same way, if we did not live in a society that made it necessary for so many people to engage in nonthinking jobs, then creativity would not be so rare. Revise educational procedures so creative thinking is encouraged, allow computers and robots to do the world's routine low-brain work, and creativity will become as common in the twenty-first century as literacy is in the twentieth.

But will there be enough creative jobs for the world's population? What kind of creative jobs would there be? How would the world's billions fill them?

That is an extraordinarily hard thing to predict in detail. Put yourself back in 1790 for a moment, when the United States was a brand-new nation with a population of about four million. Well over 90 percent of its population was engaged in farming in one way or another. Now suppose that someone, an Isaac Asimov of the past, was casting his mind into the future, predicting that the time would come when the United States would have a population of about 240 million and that only about 4 percent of them would be engaged in farming.

Naturally, the question would arise: "Then what would the other 230 million people do?"

What would the futurist of 1790 say in answer? Would he explain that some would be airline stewardesses? That some would be advertising copywriters? That some would be television actors? That some would be photographers? It's not likely that he could see the future society in accurate technological detail.

All he could say would be that there would be all sorts of nonfarming work to do—even if he lacked the ability to see the details. In the same way, there will be creative work of all kinds for people to do in the future, as robots do the scut-work and as technology continues to advance.

More and more people will be engaged in programming computers, in devising new kinds of educational programs for teaching machines. More and more people will be engaged in space and scientific research of new kinds, in technologies as yet unborn. And there'll be more teaching and acting and supervision and writing and so on.

There'll be no problem.

But how do we get from here to there? The robots that come in will replace people, and those people will not automatically become creative.

True enough. There'll be a problem. All human experience shows that, while technological advance may destroy jobs, it will also create jobs and that, in the long run, the jobs created are far nore numerous than those destroyed.

However, individuals can't always await the long run. In the near future, as computers and robots take hold, it will be necessary to initiate retraining and re-education programs. It may be necessary to have public works projects established. In short, there will be a transition period in which, in the interest of social stability, those who find themselves without a means of making a living will nevertheless be helped to find a way of living meaningfully and with dignity.

This will be expensive, and the pangs of transition must be distributed to some extent through the population. No one can rightfully expect to be untouched while many millions suffer dislocations. In the end, however, such sacrifices will be beneficial even to the sacrificer. Why? A resentful and miserable underclass is ripe for violent revolution.

The sacrifices will be temporary, especially if governments institute wise and humane policies. Eventually, perhaps in the space of a generation, new modes of education will create a population actually *predisposed* to creativity, both eager and able to take advantage of the new technology.

But what of the future beyond the next century? Surely computers and robots will become ever more complex and versatile, ever more capable of approaching the human way of thinking. Might they not then prove able to do more and more jobs, develop a creativity of their own, displace human beings from every niche, making us obsolete?

That might be so—if human beings are unable to recognize this danger and are incapable of preventing themselves from creating their own replacements.

I don't think that is likely. We *can* be aware of the danger. To be sure, human beings have allowed the production of nuclear weapons, which, in their way, can make human beings obsolete. That, however, was under the stress of national rivalry and with an imperfect sense of the true danger. With each year, as our understanding of the consequences grows, the opposition to such weapons is intensifying. Twenty years ago, world public opinion forced an end to most atmospheric testing, and in 1987, for the first time, some nuclear weapons were eliminated. I think that the danger of robots that are too human would arouse fears even more intimate and personal than nuclear weapons have. The danger would be recognized soon, and objections would come rapidly and loudly. The danger, in fact, is that human beings will *overreact* and cry for an end to robot development when, in actual fact, this research represents no harm.

Besides, in thinking that computers and robots will replace human beings and make us obsolete, we are making the assumption that there is only one kind of intelligence.

Computers, at the present moment, are far more intelligent than human beings, if we measure intelligence by the ability to solve mathematical problems. The cheapest pocket-computer can outmultiply and outdivide even a very clever human being.

But we *don't* measure intelligence in that way.

Intelligence is something much more subtle, which we can't define very easily. There will be every tendency to develop computers and robots in ways that will improve *their* variety of intelligence, rather than try to bend them in the direction of *our* variety. The analogy is to the automobile, which we try to make better and better as a thing moving on wheels. We never aspire to develop one that will move on legs.

In short, in computers and robots we will have a second variety of intelligence, which, combined with our own, will face the Universe more surely and to greater effect than either could alone.

# 3

# The Global Computerized Library

The progress of human knowledge might be coming to a halt through its own superlative success. We have learned so much, it is becoming difficult to find the specific items we need among the vast mass of the whole, specific items that may be crucial to further advance.

The sum total of human knowledge lacks an efficient index.

How can we correct this but by calling on a more-than-human memory to serve as an index, and a faster-than-human system of retrieval to make use of the index? In short, we need a computer.

Assuming that our civilization continues to exist and advance its technology, the computerization of libraries is inevitable. More and more information will be recorded on microfilm, and more and more of that will be accessible by computer.

There will be a tendency to centralize library information so that a request for particular items can tap the resources of all the libraries of a region or of a nation, by way of interconnected computers.

The process is bound to be a gradual one, so it is difficult to set an exact time for this crucial change-over. But surely, in fifty years at the most, the process will be well-advanced.

Each nation or region will have, by then, a national or regional computerized library, which will, of necessity, lead eventually to a Global Computerized Library (GCL) in which the reasonable sum total of knowledge will be stored and from which any item of that reasonable total can be retrieved on demand.

I use the word "reasonable." Although computers can handle the material once it is stored, human beings will have to decide which items to insert. There might be a desire to shovel in everything, but a sensible selectivity would surely lead to greater efficiency in the final product. There can always be subsidiary libraries which can contain outdated material, apparently useless texts, and highly specialized or arcane information unlikely to arouse much interest. It would be enough for the GCL when answering a request for such items, to be able to indicate the subsidiary library or libraries likely to be of help.

These subsidiary libraries of specialized material could closely resemble libraries of today, employing librarians in the present sense. And, of course, there will be many librarians of an entirely new kind—experts who must consider acquisitions, improve programming, maintain computers, add services.

The GCL is not likely to be a single computer, rather a set of interlocking computers located in the key cultural centers of the world. All will be equally capable at retrieving, equally capable of supplying any item, but each will do so in the language of the region.

And if a universal Lingua Terra is ever established (an amalgam of languages developing spontaneously out of the needs of increasing numbers of scholars, businessmen, and travelers dealing with each other over the world), the information can be given in that language, too.

The manner in which the GCL will be tapped is no mystery; the technique is on the way. We already have communications satellites that make it possible to connect any two points on the globe in a matter of fractions of a second. Present-day communications satellites depend on radio waves for interconnection, however, and the number of possible channels they make available is limited.

In future generations of such satellites, lasers making use of both visible light and ultraviolet radiation will be used for the interconnection. The wavelengths involved will be millions of times shorter than those of radio waves, so that the lasers could carry million of times as many channels.

The day will come, then, when every human being will have a specific television channel that can be tuned to a computer outlet that will be his or her connection to the gathered knowledge of the world. The tuning could take place anywhere. A portable device might be carried on the person. You can ask (for road directions, perhaps) and be answered by voice or in a typed print-out.

For more elaborate results, more elaborate devices in the home would be required. The equivalent of a television set will produce wanted material on the screen, or will reproduce it on film or paper—stock market quotations, news of the day, shopping opportunities, whole newspapers, magazines, or books.

This doesn't end publishing as a business, of course; it merely changes its form. In the changed form, it is quite likely to become more important than it is now.

Naturally, there are practical and economic questions. Can the public learn how to use the computer outlets?

Sure, just as they learned to operate automobiles and television sets.

There will be the desire to learn, and the mechanics of operation will be made steadily simpler.

And who pays? There are any number of possibilities. Computer use can be a public service paid out of general taxation. And each individual or business could be charged metered fees. Similarly, payments to writers could be flat fees for materials acquired or royalties based on numbers of retrievals.

The GCL would be essential for scholars and for research, but this would represent a minor fraction of its use—and importance. Far more important: For the first time there would be easy, private access to all knowledge for everybody. The GCL will make it easy for people to learn. And people *want* to learn.

This might seem a dubious statement in view of the widespread resistance to learning among the population now. In schools as presently constituted, however, the individual students are mass-fed certain stereotyped subjects at certain dictated speeds, without any regard for what it is the individual wishes to know or for how rapidly or slowly he or she can absorb the information. (As for adults, only recently has it been recognized that opportunity for education is as important for them as it is for children.)

What if, instead, there is a device, in a person's living quarters, that would feed information on exactly what he or she wants to know: how to build a stamp collection, how to mend fences, how to bake bread, how to make love, details on the private lives of the kings of England or the rules of football or the history of the stage? What if all of this is presented with endless patience, with endless repetition if necessary, and at a time and place of the learner's own choosing?

What if, having absorbed some of a subject, the learner asks for something more advanced, or a little to the side? What if some item in the information fires sudden interest and sends the learner off in a completely new direction?

Why not?

It is surely likely that more and more people will take this easy and natural way of satisfying their curiosity. Every human being has three pounds of brain that must be in constant use to avoid the pain of boredom, and here will be the perfect antidote—the GCL ready at any time to talk to you about what interests you.

Each person, as he is educated in his own interests, can then make contributions of his own. The man or woman who has a new thought or observation of any kind in any field can report it, and if it does not duplicate something already in the GCL, it can be held for confirmation and, possibly, be added eventually to the common store.

Each person will be a teacher as well as a learner.

With the ultimate library the ultimate teaching-machine as well, will the teacher-learner lose all desire for human interaction?

Of course not. Thc GCL cannot replace human contact in all areas. In athletics, in public speaking, in the dramatic arts, in exploration, in dancing, in love-making, no amount of bookishness will replace the practice—though the theory may improve it. People will still interact, and all the more intricately and pleasurably for knowing what they are doing. And they will learn enough by doing to want to know still more from the GCL, and, in turn, to teach still more to the GCL.

Then, too, every human being is subjected to the missionary instinct whenever matter in which he or she is devouringly interested in comes up. The chess enthusiast tries to get others to play chess. The same can be said of fishermen, dancers, historians, joggers, antique-buyers—name anything you want.

The person who probes the GCL and finds a fascination in weaving, or in the history of costumes, or in Roman coins, is very likely to make a determined effort to find others of like interest.

A world of people being educated, each in his or her own direction at his or her own speed, will be a world of unbelievable intellectual ferment and competition of interests. The most common frustration will arise out of having to choose among those interests.

And the GCL, as it grows more complex, can add to the ferment itself. By random association of materials in its store, it may well be able to suggest new fields of interest, new directions of research, even new conjectural solutions to old problems.

But wait—. With everyone free to learn as he wishes, will not almost everyone follow the tracks of trivia? Who will learn the dull, hard things that will be required to run the world?

In the computerized world of the future, however, it is precisely the really dull things that will be the province of automatic machinery, not human beings. To the human will be left those creative aspects of the mind that would come under the heading of "amusement" to those involved with them.

There will always be those who would find amusement in mathematics, in scientific research, in literature and art, in politics and business. They would help "run" the world, but out of the same kind of desire and pleasure enjoyed by those who are occupied in the building of rock gardens or the devising of gourmet recipes.

In a world of leisure amd amusement, though, might we not all fall apart? Is life going to become a universal Sunday afternoon in the suburbs?

Adventure? Risk? Danger? Where would *they* be?

Perhaps these will not be found on Earth in the future we are imagining, but neither will Earth be humanity's only home. Aided by the rapid advance in technology made possible by the GCL, space will be explored, exploited, and settled at a greater speed than might now seem possible, and it will be space that represents the new cutting edge of humanity.

Out there on the new frontier, the largest frontier we have yet seen, there will be adventure, risk, and danger enough for those who will find their pleasure in facing them! In this they will have the help of new branches of the ever-useful GCL.

It will be very difficult for our descedants to try to imagine what life was like without the GCL. How they will pity us!

# 4

# What Computers Won't Do

If we look into a peaceful future, it seems certain that computers will continue to become more capable and versatile. It is risky to predict, however, what computers *won't* be able to do, for one is far too likely to be wrong.

Arthur C. Clarke's most quotable comment is this: "When a distinguished, but elderly, scientist says that something is impossible, he is probably wrong."

My distinction is perhaps debatable, but I am as elderly as anybody and I take that statement to heart. However, I am perfectly willing to predict what a computer *won't* do, even if it *could.* But perhaps I should rephrase that and say that I am willing to predict what a computer won't be *designed* to do, even if the design is possible.

To see the difference between *can't* and *won't,* consider the automobile. It speeds along on wheels, which in turn spin on axles. The wheel-and-axle is the first invention human beings made that outdid nature, for no living organism progresses by means of wheels and axles. These may be impossible in organisms because of the difficulty of arranging a circulatory and nervous system to nourish, and control a living, turning wheel.

The result: While the automobile speeds along, we human beings are condemned to trudge—clumping along by lifting first one foot, then the other.

And yet walking—up, down, up, down—has its advantages. Wheels need a reasonably smooth surface, whereas in walking we can step over small obstructions and clamber over large ones. We can walk in underbrush, along narrow trails, sidle along precarious footholds while holding to a cliffside. Such things may not be as impressive as zooming sixty miles an hour along a smooth highway, but if you couldn't do them you would feel the restriction.

I imagine that it is possible to invent a mechanical device that would lift feet rather than turn wheels. If as much energy and ingenuity were put into such a walking machine as has been put into a rolling machine, I dare say we could have very nice walkmobiles. You could get into one and go walking along a rocky road, up a country lane, over rocks, and along cliffsides.

But who on earth would bother to design such a machine? Who would spend large sums in order to produce something that human beings could do easily for themselves. Even granting that walking is tiresome, it can at least be done "for free." To get into an undoubtedly expensive machine to do the same and have to put out money for fuel and for repairs in order to walk mechanically is a kind of conspicuous consumption that would appeal only to psychotics. As a matter of fact, society has proven much more willing to build an incredibly expensive network of highways in order to make the automobile wheel useful than it would be likely to spend on walkmobiles that might make highways unnecessary.

In short, then, automobiles are designed to do what human beings, without them, cannot do, or can do only with great difficulty. They are not designed, and never would be designed, to do what human beings could do easily and naturally without them.

How does that apply to computers?

We are most familiar with computers solving mathematical problems, carrying through the necessary operations at great speed and with almost zero chance of inaccuracy.

This is certainly something human beings cannot do. The human brain, while capable of working out mathematical problems, does so very slowly and tediously and with a distressing aptitude for arithmetical or logical errors. Therefore we welcome the computer in this respect and labor to design them to work on such things faster and faster and try to make them capable of tackling problems of ever greater complexity.

Why not? We don't need an automobile to go from New York to Chicago. We could walk the distance—but it would take us a great deal of time and effort. Better to use an automobile and design highways and signs and turnoffs and more economical engines to do the job ever better.

Nor are we abandoning anything vital in turning mathematical operations over to a computer. We would merely be changing an older, less efficient tool for a newer, more efficient one.

It is a mistake to think that allowing computers to solve problems "dehumanizes" a human being, and that before the computer human beings proudly solved problems by themselves. Not so.

The ordinary human being, even if reasonably intelligent and thoroughly educated, can scarcely do anything in mathematics on his or her own. If you don't believe that, and if you consider yourself intelligent and educated, why then, divide 72,647 by 323 to three decimal places *in your head.* I doubt that you'll even try (I wouldn't), and you could probably walk from New York to Chicago in less time than it would take for you to get the right answer. And yet that is an extremely *simple* problem.

All through history, we have solved even the simplest mathematical problems only with help—with our fingers, with pen and paper following memorized rules, with abacuses and slide rules and mechanical calculators. And now we have the computer, which is better than any of them.

Anything for which we can work out clear and complete instructions, and which human beings can do only with difficulty, if at all, will be handed over to the computer, and rightly so.

What about things, though, that human beings can do easily and yet for which it is extremely difficult to work out clear and complete instructions?

Here is a borderline case: chess.

Chess is played with thirty-two pieces of six different kinds on a board containing sixty-four squares in an eight-by-eight array. Every different chessman has a particular position at the start and can move only in certain simple ways. All the rules can be written out, and yet, despite years of effort, chess-playing computers are just getting to the point where they can match a grandmaster. A computer still cannot beat Karpov or Kasparov, let alone Bobby Fischer. It may some day, but it can't yet.

Why is that? Well, despite the fixed and simple starting positions and rules of movement on a small board, the number of possible positions and movements *in total* is unbelievably enormous, and we still can't get a computer to check all possibilities in a reasonable time.

But, then, how do the chessmasters do it? Ah, there you have the problem. We don't know! What's more, the chess masters don't know!

What if you take a more complicated game, then? The English language has hundreds of thousands of words and it may be that I have at my easy command five or ten thousand of them. Thousands of words instead of a few chessmen, and the words can be put together according to rules that are enormously more complicated than the rules governing chess moves. How about the game, then, of writing a story or an essay?

We all know the same words (assuming we are English-speaking), and we all know the rules of combination well enough, and we have all read stories and essays so often that we know what the finished products looks like. And yet few of us would even try to write a story or essay in the hope of having it printed. Of those who do make the effort, few can get an editor to agree that the result is publishable.

And yet I can do it. I have written, and published, quite literally, thousands of stories and essays, and I have further published, so far, more than four hundred books. There is only one way in which I can turn out this volume of material—and that is by writing as quickly as I can and getting it nearly right the first time. I do very little revising.

There is, you can well imagine, very little time for me to think, and any thinking I do manage to do has to be done very quickly.

Well, then, how do I do it? The answer is simple: I don't know how I do it! I haven't the faintest idea! I only know I've been able to do it since my teen-age years without being taught in any way.

In this I am not particularly remarkable. A great many people can do things that are extremely unusual. Who taught Mozart how to write symphonies? Who taught Louis Armstrong how to play the trumpet? Who taught Willie Mays how to catch a fly ball? Any human with a normal brain can do something or other very well and not be able to explain how he does it.

That is the glory of the human brain—that it can do things for which we are not yet able to write the rules. It may not be much good at mathematical operations or at graphic visualization, but it has what we might call creativity, intuition, insight, fantasy, imagination. It can consider a problem in which the data presented are insufficient for a certain conclusion, and yet it can guess, or feel, or intuit what the right answer ought to be. This is done all the time in business, in administration, in science, in literature, in art.

You might argue that this sort of creativity, this instinctive ability, this talent (or even genius, if you will) is confined to a very small fraction of the population. It certainly seems to be so confined, but is that our irrevocable destiny or only the result of the kind of life we live? Because we have spent all of human history without computers and, in fact, without an advanced technology of any kind, most human beings have been forced to spend their lives at work that does not significantly utilize the brain. They have had to do unskilled muscle work, they have had to do trivial mental work, they have had to engage in tedious occupations for which the brain isn't suited—adding up columns of figures, for instance.

Only the tiniest fraction of human beings have ever been in a position to exercise their creativity.

Could Napoleon have demonstrated his military skill if circumstances and economic necessity had forced him to be a tailor all his life?

In this respect, the computer promises to be the most humanizing invention in history. It will take from the suffering shoulders and minds of humanity all those tasks that human beings cannot do very will, and it will leave them those tasks for which the human brain is particularly designed.

But might it not be possible, as computers are designed to be more and more versatile and to become capable of learning by their mistakes, that in the end a computer might be able to take over those tasks that *are* peculiarly human?

It would be dangerous to say that computers *can't* ever do so, just as it would be dangerous to say that an automobile *can't* be designed to walk. I suspect, though, that computers *won't* ever do so.

Why should they? A computer, however expensive, is of great value if it can do what human beings can't do. Undoubtedly a computer would have to be more expensive still to be capable of doing what the human brain is particular designed to do, and who would want it when human beings can do it so much more cheaply? Would I want a computer, designed at enormous expense and always capable of "going down," simply in order to have it write stories and essays for me, when I am capable of writing them so easily for myself (using no more than pen and ink, if I have to)? Would anyone?

Computers may be a tool to help me do the writing—and I am using a word-processor right now to help me with the mechanical task of writing this essay. The word-processor forms the words, and then prints them up, much more quickly and neatly than I can possibly manage by my own hand. Still, I do the thinking just as much with the word-processor as with pen-and-ink, and there would be no more sense in having a computer think for me than there would be to have an automobile walk for me. Especially since I enjoy thinking much more than I enjoy walking.

There, then, is my rule for the future.

Computers probably can be designed, and probable *will* be designed, to do anything that the human brain finds difficult or tedious to do.

Computers probably can be designed, but almost certainly *won't* be designed, to do anything that the human brain finds easy or pleasurable to do.

# 5

# The Future of Handicraft

The handicrafts are extremely old, as old as humanity itself.

The very first organism that appeared on Earth and was sufficiently human to be placed into the category of "Homo" (Latin for "man") was *Homo habilis.* The Latin word *habilis* means "handy," or "skillful." This early ancestor of ours is, therefore, named "handy man."

He received this name because he was the first organism of any kind that seized upon the objects of his environment and systematically modified them in complex fashion in order to produce tools.

Still earlier ancestors may have used unmodified tree limbs or long bones as weapons. In fact, chimpanzees do so now. Chimpanzees even introduce very simple methods of modifying such objects at times. They will, for instance, strip leaves from twigs to make long pointed objects with which to poke into termite holes.

Homo habilis, however, crossed a line that had never before been approached when he began to work with stone, a material far more recalcitrant than wood or bone, and to chip it, split it, and flake it in order to obtain objects that could be used as scrapers, drills, and knives. These were procedures that required forethought and even deliberate artistry, something no form of life had been capable of till then.

The oldest examples of *Homo habilis* may have been 1,800,000 years old. Since then, *Homo habilis* slowly evolved to *Homo erectus* which evolved to *Homo sapiens neanderthalensis* and finally to *Homo sapiens sapiens* (sometimes called "modern man"). We can follow the development of the brain in the course of this human evolution by noting the size of the cranial capacities of ancient skeletons. We can also follow the increasing skill of their handiwork by noting the steady increase in sophistication of the stone tools found in connection with those skeletons.

In late prehistoric times, human beings began to use new materials for their work—leather, for instance, and textile fibers. They learned how to make baskets and shoes, to say nothing of clothing that was light and flexible. They learned how to make an artificial stone, easier to shape and handle than natural stone, in the form of baked clay, which also was used for pottery. In early historic times, they learned to isolate metal from their ore and fashion them into tools (and weapons) much superior in hardness, toughness, and sharpness than had been possible before. Perhaps an even greater advance: They learned how to make use of tools to help them in the making of tools—as in the case of potter's wheels and grindstones.

Until the Industrial Revolution totally changed man's way of life some two centuries ago, the operative devices were human hands. Objects were manufactured by human individuals with widely different skills and artistic instincts. No two objects of the same sort were made exactly alike, even when produced by the same artisan. Indeed, the makers of tools and ornaments deliberately added personal flourishes to each separate piece, on which they would then proudly sign their name.

It is easy to look back upon these times as a golden period when all manufactured articles were individual works of art, but we must be careful not to revere the situation unduly. Not every artisan was a true artist, and the few who were could only make so many objects in a given time. Those who were rich and powerful could have even the simplest household utensils made delightful and indeed unique in their proportions and artistry, but the large majority of people could only get lumpish objects, or none at all.

The time was coming, though, when machines would be devised to cut, press, stamp, and mold objects, turning out large quantities of individual items all exactly alike—and such a procedure could be useful. For instance, if a complicated device broke apart in the old pre-machine days, a substitute part had to be carefully fashioned to fit into the device exactly. Needless to say, the replacement part had to be made from scratch for the purpose. It was a slow process, and several attempts often had to be made before one worked.

That process—and hence the world—changed in 1798. In that year Eli Whitney obtained a contract to manufacture ten thousand muskets for the government. Whitney machined all the parts that went into the muskets with such precision that any particular part could substitute for any other of the ten thousand. The story is that in 1801 Whitney brought some of his muskets in to show a government official. He disassembled them and threw them down at the shocked man's feet. "There are your muskets," he said, and, picking out parts at random, put together a working musket.

Since then the process has been extended and refined. Today, in fact, the machines that make tools are being computerized so that they work more quickly and more exactly than ever.

This has its advantages, of course. Mass production is the only practical way of producing objects in sufficient numbers and of sufficient quality to make it possible for the bulk of the population to obtain what is needed to keep the standard of living reasonably high. The fact that the average American can obtain, without undue difficulty, anything from dishes to dishwashers, from shoes to shoe-trees, from automobiles to snowmobiles, is entirely due to the fact that craftsmen are not working, slowly and painstakingly, on these objects on an individual basis. We can acquire them because workers are turning them out by whirring, automatic machinery.

But there are two disadvantages. Those who create the objects are removed from them. There is machinery between the workman and his product, and the workman now often feels no sense of having created that product himself. What's more, even if he lumped himself and his machine together, and tried to feel that in combination they are producing something worthwhile, he often creates only part of the product. Furthermore, he is often required to do his work by endless repetition of essentially meaningless motions, the value of which he cannot readily see. Tightening a nut or fitting a piece of metal into a slot does not leave him with a sense of creating of an object.

The second disadvantage is to the consumer, who finds him- or herself possessed of an object totally devoid of personal character since it is exactly like any number of other such objects possessed by any number of other people. It is hard to maintain a feeling of individual pride in a particular possession under those circumstances.

Despite the fact that few indeed would actually want to abandon mass production and go back to a time when a thin layer of aristocricy had art and the vast bulk of the population had nothing (the odds are that you would be part of the vast bulk, you know), there remains an understandable nostalgia for such things as pride in workmanship, as individuality in product, as uniqueness in possession.

And there is fright at the thought of a future in which this swing from craftsmanship to computerized automation can only become more extreme, to a culture in which nothing will be individual and in which everything will be as replaceable as the parts of Mr. Whitney's muskets. What kind of brave new world will we be living in then?

And yet I don't think this particular nightmare is something we really need fear, for there are any number of cases in which advancing technology has minimized the need for craftsmanship without totally abolishing the craftsman.

It was thought, for instance, that recording devices that grow ever more precise and capacious would remove all need for listening to living musicians on real instruments (except for the brief time that it would take them to play into skillfully placed microphones.) In fact, with computers now capable of orchestrating any sound that living musicians can make and a great many they cannot, the fear has been that we won't even need people to make the recordings.

But that's not so. There is something about listening to and watching a live musical performance that technology cannot entirely duplicate. Even the chance that a living musician may hit a wrong note adds an element of excitement to the procedure. Furthermore, there is nothing quite like watching a live musician reacting to his own music.

You might say: Yes, but the number of bands and orchestras is far fewer than there would be if there were no recording devices. Not necessarily! If there were *only* living musicians, then their price would rise and, as in the eighteenth century, only the very rich would have their orchestras, while the general population would have only the rudest of music-making, if they had anything at all.

Again, photography might seem to have destroyed painting and portraiture, but you know it has not. Color photography can duplicate reality with precision, but the painter's eye does not wish to duplicate reality. It emphasizes some things and eliminates others, and deliberately alters still others so that in the end the artist produces something that may resemble reality but widely increases its significance. Furthermore, as photography produces a realism that artists find difficult to duplicate for a general public whose tastes are perhaps blunted by the camera, those artists move to various forms of impressionism and abstractionism. Rather than destroying them, photography liberates these artists, who no longer have to merely reproduce a scene.

And, as a matter of fact, photography in the hands of a master has become an art form of its own.

To move into another area, the fact that technology has eliminated the necessity of hard muscular effort, that it has replaced the shovel with the steam shovel, the pick with the jack hammer, the saw with the chain-saw, the walking feet with the turning wheel, does not necessarily mean that our muscles fall into disuse and wither away. They can, certainly, and in some cases do. But, to those who value their own fitness, involuntary servitude is replaced by voluntary activity. There are the joggers and runners, the calisthenicists and tennis enthusiasts. In fact, even though we all have the opportunity to watch sports professionals do things with a skill and finesse we cannot hope to match, that does not present us from stumbling through our own games for the fun of beating (and being beaten by) other bums equivalent to ourselves.

In short, craftsmanship isn't something you do only for money. It is doing something with your mind and hands that is individual and involves the production of something you can equate with yourself and make part of your personality—even if it is only for fun. In fact, it is hard to expect anything else to be as much fun. This is something you cannot easily give up and cannot help but find attractive. The fact that you may not have to do it for a living does not necessarily stop you from doing it for pleasure.

What it amounts to is that we are entering an age of leisure precisely because the world will be so computerized and automated, and there will be a premium on doing something that will give one's life meaning. Handi-crafting of one sort or another is the logical way. It may be something that is scientific or artistic—chemical research, musical composition, literary creation—the things that scientists, artists, and writers do now. It may be something that is simply constructive: woodworking, pottery, creating one's own appliances, applying personal art to any of the objects we want to be surrounded by.

For that matter, any craft, however old or passé, can serve to pass the time agreeably, can be an outlet for the artistic impulse. We can even chip and flake stone tools, if we wish.

And there will be others who, even in an age of automated perfection, will want something individual, something out-of-the-way, something with a feeling of art. If there are those who produce, there will be those who consume. In an age of automated perfection, there may well be more artists of all sorts, and more art, and more of a demand for art, than there could be in a time when there is leisure for the few only.

But there's not much excitement in just seeing a future in which we will continue doing what we have always done, just on a larger scale. Will we be doing anything *new?*

For one thing, we will be going out into space, and there will be a whole new field for handicrafts there. I don't know exactly what people can do with a vacuum, with zero gravity, or with temperature extremes, but these are tools and conditions that imaginative and artistic human beings can use to produce objects that will be entirely different, perhaps, from what can be formed on Earth.

And then there is computerization itself.

The computer is merely a skeleton. The real tissue is the *program,* and every new program must be written and devised. The program itself is something that must be crafted. It is a piece of craftsmanship.

It seems to me that the most important task of the computers of the future will be as the new source of education. Programs much more complex than anything now existing must be devised to serve as an incredibly flexible way of transmitting information, of tapping into a computerized library, of making available any book, magazine, pamphlet, or, for that matter, computer program that deals with something related to some key word or phrase. Programs will allow computers to ask and answer questions. Education will be something in which everyone can participate to the fullest, adults as well as children, and everything will depend on the programming.

Surely there will be many who will work on such programs, and there will be educational artists (if you like) who will devise ways that will be the great craftsmanship of the future: the total revolution of education, the total versatilization of the human race.

And we may advance, by craftsmanship, into an adult form of humanity. At least, looking back on history from a vantage point a century or so in the future, it may seem to us that all of human history from *Homo habilis* to the dawn of true computer craftsmanship will be but the childhood of the human race.

# 6

# The Future of Chemical Engineering

When I write of the future, I must emphasize, for the sake of my own comfort and reputation, that I do not predict what *will* be.

The affairs of humanity are so complex, its motivations so many, its circumstances so intertangled, that it is impossible for the human mind to penetrate the future for any great distance or with any great certainty. The clearest and most inevitable view of technological advance will come to nothing if there should be an all-out thermonuclear war tomorrow, and the surest prediction of the disaster that would certainly follow such a war would be meaningless if humanity drew back from the brink.

All that I can do, then, is to predict what *might* be, what conceivably *could* be. I can only select one image out of an innumerable sheaf of them and present it as something I find interesting.

And even so, I might be wrong in small ways that, in hindsight, would seem laughable.

Perhaps the most startling bit of technological prediction we have on record are Roger Bacon's clear insights, written down seven hundred years ago. In describing his view of technological marvels that might follow the advance of science, he said:

> Machines for navigation can be made without rowers so that the largest ships on rivers or seas will be moved by a single man in charge with greater velocity than if they were full of men. Also cars can be made so that without animals they will move with unbelievable rapidity.

Right on. You would think that Roger Bacon had come to the thirteenth century from the future and had actually seen motor boats and automobiles. (Yes, science fiction writers have suggested that.)

But then Bacon goes on to describe airplanes:

> Also flying machines can be constructed so that a man sits in the midst of the machine turning some engine by which artificial wings are made to beat the air like a flying bird.

Right (in essence) again, but Bacon did not appreciate the fact that birds wings exist for support *and* propulsion, and that if propulsion is supplied by "some engine" then the wings need merely exist for support and can be motionless. They don't have to beat. Leonardo da Vinci also missed this. And, indeed, it wasn't till the midnineteenth century, when George Cayley founded the science of aerodynamics, that it was possible to do away with the notion that flying machines had to have beating wings.

So, with these caveats, I will try to look into the next century of chemical engineering.

* * *

All new fields of technology will involve the large-scale manipulation of atoms and molecules in some way, and that will bring in chemical engineering—inevitably.

For instance, the twenty-first century may well see the coming of practical, controlled nuclear-fusion power. Fusion power should prove to have a number of advantages over the nuclear-fission power presently in use. Its basic fuel will be deuterium rather than uranium or plutonium, and deuterium is much more easily obtainable in quantities large enough to last human beings for as long as they are likely to remain on Earth or in existence. Fusion will not require a minimum "critical mass" of fuel as fission does, and it will be able to work with microscopic quantities, so runaway excursions won't be likely. Fusion also will not produce radioactive products in the quantities and intensities that fission does.

Fusion energy will be used for all purposes that energy is ordinarily used for, but one rather simple application may be unique to itself. We can imagine a "plasma torch," a superhot jet of gas powered by nuclear fusion, hot enough to vaporize any material and to break up any molecule into its constituent atoms.

The value of this would be that it would be a universal garbage disposal unit, so to speak. We are now living in a world in which wastes are an increasingly troublesome and even intractable problem. Ordinary physiological wastes are biodegradable and are recycled into usable forms in the biosphere. However, many chemical wastes are toxic and long-lived, and some solid wastes are not biodegradable. We have reached the point where toxic wastes threaten our water and air supplies and put at risk the very viability of the planet, while we are rapidly running out of places in which to put our ever-increasing mountains of solid wastes.

The plasma torch can consume all wastes that cannot be handled naturally and, by converting them into their elements, will make it possible to reinsert them into the various natural geological and biological cycles.

Of course, it isn't quite that easy. (After all, what is?) In torching wastes, you produce a mixture of elementary gases, some of which are corrosive and, therefore, toxic. One must imagine a torching chamber built out of inert materials, together with devices for bringing about, as quickly as possible, those chemical reactions that would nullify toxicity and corrosion. You don't want either sodium or chlorine vapors to exist for long, prefering to have them converted into sodium chloride with as little delay as possible.

Then again, it would also not be desirable to end up with some complex slag of limited use. Somehow the torching chamber should be so designed that the products are sorted out into separate batches of materials. It would be useful to end up with ferrous metals here, nonferrous metals there, and silicates yon.

Ideally, the products should be separated to the point where they could easily be purified still further and many of the products reutilized. The strain on the Earth's mineral resources would be greatly eased in this fashion. We don't actually consume the Earth's elementary resources, but we do begin with portions of the Earth where geological processes have concentrated certain elements and end with them well-mixed and hard to recover.

The plasma-torch process, properly designed by ingenious chemical engineers, would offer us a way of unmixing the elements, and using them over and over again, indefinitely. It will be at the expense of much energy, for we are enforcing a local entropy loss in unmixing the elements. But, with fusion power in existence, that energy should be easily available.

In fact, I dream of one particular application of the plasma torch that may be of greater importance than anything else.,

Throughout history, since the discovery of fire, human beings have obtained the lion's share of the energy they have used by burning wood, coal, oil, or gas. In every case, this has meant the combination of organic molecules, containing carbon and hydrogen atoms, with oxygen to form carbon dioxide and water.

This is not a drastically dangerous process in itself, for the plant world recombines carbon dioxide and water to form organic molecules and oxygen and thus maintains the balance (at the expense of the energy of sunlight).

In the course of the present century, however, the burning of fuel has proceeded at a rate that has outpaced the ability of the plant world to restore matters. The tiny quantity of carbon dioxide naturally present in the atmosphere (about 0.03 percent) has been slowly creeping upward.

The greater supply of carbon dioxide in the air does not interfere with our breathing in the least. But carbon dioxide does tend to be opaque to infrared radiation, which means that it interferes with the Earth's loss of heat to outer space at night. In other words, the additional carbon dioxide brings about a slight rise in the average temperature of Earth. This can eventually lead to the melting of the icecaps and a marked (and unfavorable) change of Earth's climate. (This is called the "greenhouse effect.")

If we have fusion power, however, the use of fossil fuels will surely decline. In addition, might it not be possible to pass air steadily through some torch chambers, breaking up carbon dioxide molecules, liberating the oxygen content, and gradually accumulating carbon? We would, in effect, be remanufacturing coal, and it would be the task of chemical engineers to make the process as rapid and as efficient as possible.

It is not at all likely that we could, in this way, withdraw carbon dioxide as fast as it is being pumped into the atmosphere today, but in a society in which carbon dioxide is being formed only minimally, the implementation of such torch chambers might be enough to help reverse the greenhouse effect.

* * *

Chemical engineering will meet a still greater challenge in the next great extension of the human range.

In the second half of the twentieth century, humanity made its first incursions into space, sending astronauts to the Moon and uncrewed probes to the surface of Mars and Venus as well as to the immediate vicinities of Mercury, Comet Halley, Jupiter, Saturn, and Uranus. (Before the end of 1989, *Voyager 2* will have sent back information from Neptune as well.)

In the twenty-first century, if the nations of Earth can reduce their mutual hatreds and suspicions and learn to cooperate in gigantic nonmilitary projects, it should be possible to advance into space in a *permanent* fashion, to colonize as well as explore it.

The first step will undoubtedly be space stations with permanent crews aboard (in shifts, of course), who can, among other things, carry through the building of various structures in space. These might include solar power stations, observatories, laboratories, factories, and, most of all, settlements, each of which might become the more-or-less permanent home of ten thousand human beings.

Naturally, it is difficult to imagine the materials for these structures being drawn from Earth's already strained resources and for it all to be brought from Earth's surface into space by the main force of chemical-fuel rockets.

It may well be, then, that the next major step, after space stations, will be the establishment of mining stations of the Moon. Human beings will return to the satellite to stay.

The Moon's surface is, at present, utterly unused. It is a completely lifeless world, so that not even our noblest idealists can argue that it belongs to native life forms. It is a large world, with a surface equal to the areas of North and South America combined, and yet it is considerably smaller than the Earth itself, so that its escape velocity is only 1.5 miles per second—as compared with Earth's 7 miles per second.

This means that material from the Moon's surface can be hurled into space with the expenditure of far less energy than would be possible from the Earth's surface. The Moon would also benefit, in this respect, from being without an atmosphere to supply air resistance and the viscissitudes of weather.

Indeed, a great many theoretical studies have been made to show that it is practical to suppose that quantities of the Moon's surface materials can be hurled into space by electromagnetic propulsion—these so-called "mass-drivers" being powered by solar energy, which would be ample on the Moon's surface.

The lunar material could be smelted in space to yield the various structural metals, aluminum, iron, tungsten, titanium, and so on. Treated in other ways, the material could yield cement, concrete, glass, and oxygen. The only badly needed materials that the Moon lacks are carbon, hydrogen, and nitrogen, and these the Earth could supply without trouble until alternate supplies become available.

In smelting and otherwise treating lunar materials, however, we must have large scale chemical processes in space, and it would be up to chemical engineers to design these. Undoubtedly, these designs would be notably different from those of similar structures on Earth since in space there would be essentially zero-gravity, endless vacuum, and the hard radiation from the Sun.

This would hold true for all other structures in space, each of which would supply chemical engineers with new problems.

The solar power stations, for instance, would require square miles of photovoltaic cells capable of converting light to electricity—and these would have to be formed in space. Smelting and welding in space (thanks to its hard vacuum, which on Earth could be made to exist only temporarily, over a small volume, and at the cost of much energy expenditure) might result in substances freer of impurities. The result could well be cells that are longer-lasting, cheaper, and more efficient. We could not judge the value of solar power stations in space until we find out just how well chemical engineers can take advantage of the unusual properties of space to produce the necessary materials.

Naturally, space has its disadvantages, too. It is a dusty place, with each particle of grit traveling miles per second, so collisions will make their mark. (Indeed, already the human incursion into space has produced large numbers of dead satellites and parts of satellites, right down to the existence of innumerable flecks of speeding paint and rust, all of which offers the chance of damage.) It should be a major portion of the chemical engineering art to so design the cells as to minimize the damage done by colliding supra-atomic particles, or, for that matter, by the impact of energetic charged subatomic particles of the "solar wind" that is always issuing from the Sun.

An even more enormous job will be that of designing factories in space. Until now, our industrial plant has, perforce, existed on the surface of the Earth. That means that any dangers that might accompany industrialization (fires, explosions, and so on) must take place in the midst of earthly life. Even if industries are far removed from human population centers, they can still do damage to vegetation, to soil, water, and atmosphere, on all of which, of course, human life depends. Indeed, chemical wastes, in particular, threaten to poison us all. Even small amounts of chemicals can have terrifying consequences; we need only note the conversion of industrial smoke into acid rain, and the decimating effect chlorofluorocarbons have on the ozone layer.

Naturally, we cannot reasonably hope to abandon our industries and "return to nature." Unfortunately, the population of the Earth has risen to five billion. Before the world had industrialized itself, however, it was able to support not quite one billion people, and these not at a very high standard of living. If we were to deliberately deindustrialize now, we would be doing the equivalent of telling some four billion people to get off the Earth, and in the confusion of trying to decide which four billion it would be, all of us might die.

The reasonable alternative is to move as much of our industrial plant as possible away from Earth's biosphere; that is, off Earth and into orbit. In this way, industry will be gone (or, at least, partly gone) from our midst, yet it would not disappear altogether. It would merely be a few thousand miles away—straight up.

We can envision factories that will be largely automated and roboticized, supervised for the most part from long distance (as our rocket probes are), and rarely requiring actual on-the-spot human attention.

Many have speculated on the possibilities of new advances in technology resulting from the unusual properties of space. The absence of perceptible gravitational effects could make it possible to produce more nearly perfect ball bearings, for instance. The presence of hard radiation from the Sun (radiation that does not reach Earth's surface because of the filtering effect of the atmosphere) may make some processes (such as photochemical reactions) easier, and others (such as maintaining the stability of complex molecules) more difficult.

However, all the possibilities will require hard and fast processes on a large scale before they become anything more than that. One can envisage "orbital chemical engineering" as a new and spacious branch of the field, in which engineers will be dealing with entirely new situations, facing new problems. And we can surely hope, coming up with new solutions that will make the twenty-first century as different from the twentieth as the twentieth is from Roger Bacon's thirteenth.

* * *

But if orbital chemical engineering promises to be one of the glamour subjects of the twenty-first century, biochemical engineering will surely touch us even more closely.

Of the molecules that chemists and chemical engineers must deal with, none are so complex and, on occasion, delicate as those found in living tissue. And of these "organic molecules," none are as complex and, on occasion, delicate as proteins and nucleic acids.

Each of these two types of molecules are polymers, long chains built up of relatively small units that repeat themselves along the chain from dozens to thousands of times. In the case of proteins, the units are some twenty different amino acids; in the case of the nucleic acids, four different nucleotides.

If we consider the proteins first, each amino acid is build up of a chain of three atoms, one nitrogen and two carbons (N-C-C). To the middle carbon is attached a side-chain, and the side-chain of each amino acid is different. Some side-chains are small, some large. Some carry an electric charge, some do not. And where there is an electric charge, some are positive, some negative.

Once the amino acids string together, they fold into a three-dimensional molecule, and the amino-acid side chains form a lumpy, uneven surface with electric charges of both types scattered here and there. Every different arrangement of amino acids produces a surface of its own characteristic shape, and the number of different arrangements is inconceivable.

If you begin with only one of each of the twenty different kinds of amino acids, you could line them up in more than $2.4 \times 10^{18}$ arrangements, each producing a molecule of a slightly different shape.

Actual protein molecules, however, consist of far more than twenty amino acids, and with widely varying numbers of each. The hemoglobin molecule can have its amino acids lined up in $10^{640}$ arrangements (a number that is so great there is nothing in the material universe with which it can begin to be compared), and only one of these arrangements will work perfectly.

Some proteins are structural substances, making up hair, skin, and connective tissue. The most important ones are enzymes—that is, catalysts. Each enzyme has a characteristic surface of a shape such that a particular small molecule will fit neatly onto a certain portion of its surface. The small molecule combines with the enzyme and is held in place in such a way that it can easily combine with another molecule and undergo a chemical change—a change that would take place very slowly or not at all if the molecules were not on the surface of an enzyme. Once the chemical change has taken place, the reaction products no longer fit the surface and are released—making room for another molecule.

Every living cell has thousands of different enzymes, each capable of bringing about some chemical reaction. With all these enzymes present, the cell is a hive of rapid, interlocking reactions, all of which, together, maintain the chemistry of normal life.

The number of possible enzymes is enormous. It is because of the different combinations of these enzymes that one species is different from another and that some ten to twenty million species have evolved on Earth through the eons. It is because of these enzyme differences that two million different species still exist today and that millions more may evolve in the future. It is because of minor differences in enzymes and enzyme-combinations that each individual in a species is different and that no two human beings are quite alike (leaving out of account identical twins).

The characteristic enzymes of a species (and of an individual) are produced because each cell contains molecules of deoxyribonucleic acid (DNA), which have the ability to produce exact replicas of themselves every time a cell divides. The order of arrangement of the nucleotides in different portions ("genes") of such DNA molecules contains the information for the order of arrangements of amino acids in the enzymes. Each gene guides the production of one particular enzyme.

The enzymes are much more complicated than they have to be to perform their catalytic function. It is the active portion of the surface, a region made up of a few amino acids, that does the work. The rest of the possibly enormous molecule makes certain that only one type of molecule will fit its surface, that the enzyme itself fits in with all the other enzymes or that it does its work in coordination with the entire system.

It is possible, however, to devise a small molecule that mimics the shape of the active center of the enzyme. Such a small molecule may duplicate the catalytic activity of the enzyme. The small molecule may not work with the speed of an enzyme, and it may not be as delicately selective as an enzyme. On the other hand, it would be more stable than an enzyme, and could endure rougher handling.

In 1987, three chemists, Donald J. Cram, Charles J. Pedersen, and Jean Marie Lehn, received the Nobel Prize for their work on these enzyme-mimicking molecules. It remains to apply them to large scale, commercial, catalytic processes, and this of course will be the task of chemical engineers. They will devise dozens, perhaps hundreds of large-scale reactions, using organic catalysts in place of the metallic powders and acids that are often used today. The processes could be gentler, faster, and cheaper than those used today, and there could be many that will speed along catalytically that have *no* appropriate catalysts today.

More delicately still, one can reckon with the genes themselves. Chemists are learning to map the location of the genes along the enormously long DNA molecules that make up the individual chromosomes in a cell nucleus. Particular physical characteristics can be associated to particular genes in particular locations. More important, particular inborn, metabolic diseases can be attributed to certain defective genes. Cancerous changes can be blamed on certain "oncogenes" located in certain places.

Increasingly, as more and more is learned about these genes, it becomes possible to tell, while a child is still in the fetal stage, whether certain deficiencies exist.

Chemists are also learning to determine the nature and order of the nucleotides as they occur in a gene so that, in effect, they learn the entire structure, atom by atom. This opens up the possibility of finding out not only that a particular gene is defective but that the defect lies in certain missing, or superfluous, or disarranged, or misshapen nucleotides. That, in turn, makes it possible to suppose that the time will come when genes can be corrected or altered and a condition possibly relieved that is otherwise incurable.

It might seem that this will not lend itself to chemical engineering methods; that a defective gene must be corrected delicately, one at a time so to speak; and that it must be adjusted to the individual.

However, one can imagine a more general situation.

The number of different genes that exist in all the cells of all the living organisms that have ever lived is simply enormous. Unquestionably, you have genes not quite like those in any other organism, and the same is true of me, him, her, the nearby rabbit, the oak tree, and a blade of grass.

Yet all the different genes that do exist and that have existed and that have functioned well in one organism or another amount to a tiny submicroscopic number compared with all the genes that *could* exist. (Think of the hemoglobin molecule.)

To be sure, it may be that most genes, if we design them by putting nucleotides together at random, will produce protein molecules that have no enzymatic properties. Nevertheless, there have existed enough that *have* worked to have produced millions of species and uncounted quadrillions of individuals all different, even if, sometimes, only in tiny ways.

It follows, then, that there must be an enormous number of genes that would produce enzymes that, under one condition or another, might work and be useful, that have never yet appeared—or that have appeared briefly and died out too rapidly to make their mark in evolution.

It would be interesting to try to construct some of these genes. What might the enzymes form? What reactions, if any, would those enzymes catalyze? Obviously, there is no chance that we will ever to able to study all possible genes. If every atom in the Universe were a research scientist and if each one could form and study a new nucleotide arrangement every second through the entire existence of the Universe so far, the number that would have been produced would be nearly nothing compared to the total number of genes that could exist.

Studying genes, however, might eventually teach us which groups of those substances have a greater chance of working. We could mine the richer lodes, so to speak, and study not only the individual genes but varying arrangements of genes. (Genes affect one another and the order counts, too, which makes matters still more complicated.)

People might worry about the danger of producing a gene which would be, in one way or another, dangerous, that would "take over" and create a disease or abnormality that could never again be eradicated. Actually, the chances of any danger resulting from a single gene is nil, and we must remember that new ones are, in any case, arising constantly through the processes of natural mutation.

Besides, more and more, this sort of thing might be reduced to computer simulation. A gene can be constructed on the computer screen and be converted into an enzyme according to the known rules of such conversion. The enzyme surface can be studied three-dimensionally in order to determine a molecule's fit or lack of fit.

It might be possible to learn enough about genes and enzymes to actually create new species in the computer, to work out the physical appearance and the biochemical and physiological properties of an organism that will have a certain set of genes, some or all of which don't exist in reality, or which exist but have never been arranged in that particular order before.

In this may, we might computerize evolution and follow down its possible tracks in directions that seem to us to be useful.

So far, evolution has continued for some three and a half billion years, since the formation of the earliest primitive cells in a more or less hit-

or-miss fashion. Mutations have occurred randomly, and the forces of natural selection have chosen among them and driven life along the numerous branches and paths it has taken . . . but only as far as the available mutations have allowed.

Perhaps this is the process that will continue until a point is reached where some form of life is intelligent enough and technologically advanced enough to guide evolution knowingly over a desired path, to *choose* mutations with foresighted intention. (The path may be ill-chosen and end in disaster, but surely this is not inevitable. There is at least a chance we may be wise and do fairly well.)

This is comparable to something that has actually happened in human history. For millions of years after hominids first came to stand upright on their hind legs, they slowly advanced in technology by trial and error. Then, eventually, when modern science was developed, humanity was able to advance knowingly along chosen technological paths. It took nearly two million years, perhaps, to advance from the first chipped stone to the steam engine, and then only two hundred years to advance from the steam-engine to a rocket-ship resting on the Moon. We can wonder whether we have not foolishly followed a path to destruction with nuclear weapons, but surely this is not *forced* on us. We can still choose to be sane.

In the same way, if someone looks back on history from the vantage point of the far future, he or she might see the human species finally reaching adulthood in this, our present lifetime, only now beginning its advance though sunny uplands, uplands we can scarcely see today.

But where does chemical engineering play its part in this scenario? Surely, it will be a long time before we reach the stage of computerizing evolution and studying it on some three-dimensional screen. Before we can reach that point, we must learn a great deal more about genes.

We must increasingly convert the analysis of genes, the synthesis of new genes, the conversion of genes to enzymes, and the detailed study of the functioning of new and potential enzymes to large-scale studies. This is the biochemical engineering I spoke of.

I haven't the faintest notion as to how this all will be done, but then I am not a chemical engineer. I trust that in the days to come, chemical engineers will devise methods for genetic studies on a large scale and lay the groundwork for its computerization.

The grand vision I have drawn might not be reached by the end of the twenty-first century (indeed, in my cautious heart, I feel certain it won't be). But a beginning will at least be made, and elementary biochemical engineering will be healthily on the way.

# 7
# Men and Marriage

Marriage is based on the fact that there are two human sexes that are sharply differentiated biologically, as much so today as at any time in the past.

In the first place, women bear children and men don't. This meant that, when humanity lived completely off the land, women periodically underwent nine months of pregnancy, during the latter third of which they were less agile and more vulnerable than they would otherwise be. Pregnancy and lactation are so energy-consuming that women, when compared to men, were more likely to require protection from the outside world.

Another biological difference, as true today as ever, is that men, on the average (though not in every specific case), are larger and stronger than women. This, together with freedom from pregnancy and lactation, made it possible for a man to extend protection to a woman and to undertake the task of seeing to it that she could bear her children in relative comfort and peace.

This protection was not extended for nothing! In exchange, a man could reasonably expect to have a convenient subject for sexual contact and a supply of children who could help him in the fields and care for him in his old age.

All this enforced a division of labor. The woman, tied down by children to whom, thanks to pregnancy, labor, and lactation, she felt more emotionally bound than the man did, would naturally tend to confine her activity to the home. The man, who must supply food and protection, would tend to venture afield, to hunt or to farm.

This is not a biological difference, but a social one that was encouraged by a biological difference. The division of labor need not be sharp. There was nothing to prevent a woman from sharing the labor on the farm or a man from sharing the labor in the home, and this may well have been done. But even today, the *traditional* family system puts the woman in the home and the man in the world outside. And tradition, of course, tends to be revered beyond its deserts.

In forming a partnership, a woman was selling her sexual desirability and fertility and a man his strength and potency. It was, however, more of a buyer's market for women since a man could, after all, take sex by force, if he was of a mind to, and give nothing in exchange. It fell to the woman, therefore, to do most of the advertising. Even today, it tends to be women who wear clothes designed to reveal and emphasize sexually desirable characteristics, and it is women who artificially color their faces.

There is still another biological difference that can't be effaced. Women can bear babies only while relatively young. After forty, it becomes first difficult and then impossible to have a child. Men, however, can father children even into old age.

This meant that a woman had to emphasize her youth as much as possible, so that even today it is a woman who is most likely to dye her hair, work to keep her skin taut, and so on. And by tradition, she was expected to lie about her age.

Still another biological difference is that a woman *knows* that a child is hers since it emerges from her womb. A man does not have the same certainty. If a man wished to be reasonably sure that a woman's child is his as well, he would probably want to make sure that his wife was not available to other men. Wives were likely to be kept in seclusion and under close guard for that reason.

A double standard is the natural consequence. Men could philander freely because no woman is, as a result, inflicted with a child that is not her own. Women were *not* allowed to be unfaithful, however, because that could bestow upon her husband a child that was not his. For the same reason, polygamy (men with multiple wives) is not uncommon in tiny societies, while polyandry (women with multiple husbands) is quite rare.

And how have things changed today? The biological differences remain. The old traditions based on those differences remain. Yet surely many factors in society have changed.

For instance, babies are no longer as important as they once were. In early times, infant mortality was so very high that women were virtually forced to have many children in order to guarantee that some would live long enough to be a help and comfort to their parents. Nowadays, infant mortality is quite low in advanced portions of the globe, and two or three children are quite enough to ensure a continuing, intact family under normal conditions.

In early times, too, a woman desired babies since this was virtually the only contribution she was allowed to make to society. If she had no children, she was useless both to her husband and to her community. Furthermore, the blame for lack of children was placed squarely on the woman. The husband supplied the seed and the woman was the soil in which the seed grew. If the seed did not grow, it was because the soil was barren, and that's what women who did not have children, despite having sex, were called—"barren."

It was easily assumed that barren women were cursed because of sins they had committed, which made it even worse for them. In the Bible, Sarah and Rachel did not have children for a long time and were miserable as a result, fearing rejection. The mother of Samson and the mother of Samuel were barren for a while and were in misery. And when Michal, who was King David's first wife, was impudent to her husband, she was punished by being made barren all her life.

Nowadays it is recognized that man and woman contribute equally to the child, and that failure to have children is as likely to be due to some shortcoming in the man as in the woman. Moreover, it is a *biological* failure, not a moral one.

In fact, since there are now five billion people on Earth, making overpopulation a great danger to us all, families in many parts of the world are being encouraged to have fewer children.

All these changes mean that it has become completely acceptable for a woman to have few children, or even none, if she so wishes or if things so order themselves. (The *traditional* family remains one with many children, of course, but tradition never catches up with change.) That, in turn, means that women have time to play other roles in society. Indeed, you might turn that around: women should be not only allowed to play roles other than "mother" in society, they should be *encouraged* to do so. Otherwise, they might be forced to rely on procreation as a way of attaining prestige, and we can no longer afford a high birth rate.

Then, too, parts of the double standard have lifted. It has always been possible for a young unmarried man to be sexually active, since he could not suffer pregnancy as a result and, if he were sufficiently unscrupulous, he need not worry about the woman's pregnancy. A young unmarried woman, however, had to risk pregnancy with each sexual interlude conducted without any contraceptive and, if she were pregnant, she had to bear the full blame and obloquy. (A young man, after all, was *expected* "to sow his wild oats.")

Nowadays, however, we have "the pill," as well as other methods whereby a woman can protect herself against pregnancy. The result: She can be sexually active with relative impunity. She has, in fact, become so, and social disapproval has weakened considerably. It has reached the point where a young unmarried woman can have a child without being ostracized. (However, our society is so ordered that a single woman with a child receives little economic help, so that she is usually condemned to poverty—as is her child.)

Because of social change, therefore, a woman can obtain work more easily, and with a greater choice. Gone are the days when the only alternative to starvation for single women was such jobs as companions, governesses, seamstresses, and maids, jobs that paid her exploitively low wages and in which she was treated like dirt. Nowadays, if she can survive the hatred of her male fellow-workers, a woman can be a lawyer or a firewoman or a sanitation woman.

Furthermore, she can be sexually active while avoiding children. The result is that there is not quite the pressure on her to get married. In older days, a woman without a husband could face only poverty and the disgrace of being an "old maid." A woman had a strong desire, therefore, to marry someone, anyone, simply in desperation. But today a woman can be choosy and marry someone she wishes to live with, remaining single without disgrace until she does.

Naturally, then, the traditional till-death-do-us-part marriage has become an endangered species. After all, until a century and a half ago, the life span, even in advanced societies, was about thirty-five years, so that the average marriage was ended by the death of one partner or the other after ten or fifteen years. It is not very difficult for a marriage to endure for that length of time.

These days people have an average life span of up to seventy-five years, and marriages are quite likely to last half a century if divorce does not intervene. Since the thought of fifty years with a person whose flaws become all too manifest after five is a hard thing to face, divorce has become common and can be indulged in without much social disapproval.

This is hard on children who must live in a "broken home," but, in past times, broken homes were just as common, even though they were broken by death rather than divorce. And it is well understood that children suffer as much, or perhaps more, when they are part of a family in which the mother and father live together in mutual hatred because they cannot, for religious or economic reasons, divorce.

Where does this leave men where marriage is concerned? My own feeling is that the liberation of women liberates men as well.

In the old days, women were routinely not educated. (What, indeed, did they need an education for?) It was also widely assumed that women's brains were biologically far inferior to men's. As a result, a man assumed that he *must* be brighter than his wife and would be unbearably humiliated if she showed signs of being brighter than he. That was the origin of the thought (which is still prevalent today, thanks to tradition) that a young woman must hide her intelligence and pretend to be silly or even stupid or "no man will look at her." Silliness in a woman is usually considered "cute." And, of course, if a woman never uses her brain out of policy, she eventually ceases to be able to use it at all.

This sort of thing meant that a man was supposed to live with a stupid woman. Sex quickly fades in desirability when it is a customary activity, and companionship with a fool is no pleasure. In the not-very-long run, then, a husband and wife grew tired of each other and lived lives of what Thoreau called "quiet desperation," or they got divorced.

Nowadays, women *are* educated, and a man can expect his wife to be as intelligent as he himself is and, in some ways, even more intelligent.

If he can break himself of the traditional reverence for female stupidity, he will have a much better companion and appreciate her for a far longer time. (Mental compatibility lasts longer than physical compatability and is, in the long run, more pleasurable.) A woman will in turn be more satisfied with a husband who does not mistrust her intelligence.

In short, marriage can become a true partnership, revealing delights that are hidden to most who live traditionally. Indeed, such a "new" marriage could do more to stabilize the family than the traditional marriage ever could.

If a woman takes on more responsibility outside the home, it follows that a man is under pressure to take on more responsibility inside the home. This naturally meets with resistance! Household chores tend to be dull and wearisome (which is precisely why men insisted on leaving them to women); but a man can often enjoy cooking, for instance, if he does not consider this to disgrace him. He even gets the chance of cooking something the way *he* likes it, rather than being condemned forever to eat what his wife prepares for *her* delectation.

Other chores are made easier when shared—which has the additional virtue of intensifying the bond between a husband and wife. (I hasten to say that I am not a good example of this. My wife, a retired psychiatrist and now a full-time writer in her own right, does most—but not all—of the housework. But then, as a "prolific" writer, I keep a seventy-hour week, and she understands this. The day she catches me with a can of beer watching a football game will be the day she hands me the vacuum cleaner.)

Then, too, a woman's greater activity outside the home makes it necessary for a father to share in the task of caring for the children. Why not? All the better. He forms a closer bond and forfeits the traditional role of being a remote ogre. ("Wait till your father comes home. *He* will punish you.")

A man might view this in a selfish light. Why should he not have the pleasures of intimacy with his children? Why should his wife have all the fun? Indeed, if the family should break up, he would have a better case for visitation rights and have an equal chance perhaps at attaining custody.

To put it as briefly as possible, a marriage between equals is worth far more than a marriage between two disparate and contentious nonpartners. We are fortunate enough to live at a time when, and in a society where, this is now possible. We should count our blessings and fight to keep this possibility open in the future.

# Part II
# Space

---

The nine essays in this section also deal with the future, more or less, but it is with the future as it particularly involves space. Most of these pieces were written after the *Challenger* disaster in January 1986, when many people's hopes for space passed through a mournful period of pessimism. I remained determinedly optimistic, however, and the smashingly successful trip of the *Discovery* lifted all our spirits; I felt very much on the right track.

However, optimism can miss the mark, too. My essay "All Aboard for Phobos," which dealt with the two Soviet probes to Mars's satellites, was cut in half (so to speak) soon after it was published, when one of the probes stopped working, and then altogether when the other did, too.

"Other Intelligent Life?" (this book's other moderately long essay) might also be overly optimistic. Since it was published, I have sensed a certain diminution of enthusiasm among astronomers concerning the possibility of extraterrestrial civilizations. However, I cling to the hope that they are there, and so I am including the essay in the book, perhaps as a small gesture of defiance.

The essay in this section that I'm proudest of is "The Telephone in Space." When I was asked to take up the topic, I responded, "A *telephone* in space? I never heard of such a thing and have never thought about it." The telephone people who wanted the article were obdurate and insistent, however, so I sat down to think about it and the essay I include here was the result. I was quite pleased with what my imagination produced after the appropriate flogging.

# 8

# The Lure of Exploration

The first humanlike creatures lived in southern Africa some four million years ago. Little by little those creatures and their descendents spread out over the face of the Earth. They did so very slowly at first, and then faster and faster, until now every continent (even Antarctica), every island (even the most isolated), has felt the tread of human feet.

Nor is it the dry land alone that human beings have traversed. They have plunged beneath the ocean surface, to the very deepest abyss. They have soared through the air, and beyond. Twelve human beings have walked on the Moon.

What is this lure of exploration? What drives human beings farther and farther into the unknown?

At first, it was nothing mysterious: People were searching for food and water. Droughts or population increases would make it hard for some to find and gather enough to eat and drink, so they would have to wander off in search of what they needed.

Or it might be fear. Predatory animals might invade or, worse than that, other human beings might come seeking food and water. The human beings already present would be driven off and would have to find new homes.

But whatever it was, there must also have been the excitement of new lands, the beauty of new vistas, the delight of finding a new place that could be made into a new home.

And so, little by little, the Siberian mammoth-hunters followed the migrating herds across the Bering Strait (dry land, then) and entered the American continents about 25,000 years ago, and then they penetrated ever southward until there were human beings in Tierra del Fuego, the island off the southern tip of South America. On the other side of the world, human beings made their way from island to island of the vast Indonesian archipelago until they reached Australia and, beyond it, the distant islands of Tasmania. Later still, human beings in primitive ships dared the lonely, empty reaches of the Pacific, Earth's largest ocean, and populated the large island of New Zealand and all the tiny islands that were so widely scattered over its expanse.

We know almost nothing about those epics of exploration except that they were carried out successfully. We don't know what bold individuals led the exploration teams, what dangers were dared, what difficulties were overcome, what deaths were suffered—but dangers, difficulties, and deaths there must surely have been.

Even in the more recent history of Europe and the Middle East, which we know far better, we know little of the early explorations. It was the Phoenicians, who lived on the coast of what is now Lebanon, who were the first sea-explorers of historic times, about 1000 B.C. Their mariners learned to navigate by the stars and could thus move out of sight of land (an early example of how developing knowledge increases safety). Their ships roamed the full length of the Mediterranean and emerged into the Atlantic Ocean. Searching for new sources of needed materials, the Phoenicians located the Tin Isles (off the coast of what is now England)—tin being essential for the smelting of bronze. There are even tales that in the sixth century B.C. a Phoenician expedition sailed all around Africa in a journey that lasted three years.

Nearly two thousand years later, the Vikings of Scandinavia, from A.D. 800 onward, poured out of their harsh, winter-ridden peninsula, raiding the coasts of Europe and settling in the British Isles, in northern France, in southern Italy, and along the river routes of Russia. They found new lands, too—Iceland, Greenland, even the northeastern coasts of North America.

However, the Golden Age of Exploration began in the fifteenth century, with the little nation of Portugal leading the way. European vessels, making full use of the mariner's compass, could navigate even when the skies were cloud-covered and when neither Sun nor stars were visible to give an idea of directions. They criss-crossed the oceans, established new nations of European heritage in the Americas, Australia, and southern Africa, and came to dominate the already-peopled lands of Africa and Asia. For a period of four and a half centuries the Earth became more and more European. It is only in our own times that the submerged nations have broken free again.

The driving force behind explorations was still very practical. Europe needed silk, sugar, spices, and other commodities from the East. The land routes were difficult, however, and, in addition, they were blocked by Muslim lands at enmity with Europe. What was needed was a sea-route direct to the Orient that would bypass the Muslim lands.

In the process, heroic voyages were made and dangers were met with resolute strength. Indeed, there were new dangers, aside from the old ones of storms and doldrums.

In 1497 the Portuguese explorer Vasca da Gama set out on a journey that in nine months took him around Africa to India—the first European to reach Asia by sea. During the voyage, however, the sailors fell ill with scurvy. This was the result of a diet that lacked the vitamin C present in fresh fruits and vegetables, but no one know that at the time. For centuries, scurvy continued to be the scourge of long voyages, and many sailors succumbed to it.

In 1519 the Portuguese mariner Ferdinand Magellan (in the pay of Spain) set out with five ships and 230 men to reach the Far East by travelling westward. In circumnavigating the world this first time, these men endured a stormy passage through the Strait of Magellan and the ninety-nine successive days of crossing the Pacific without any sight of land; they all nearly starved when their food ran out. While they brought back a load of spices that made the voyage profitable, the cost was heavy. Two hundred and twelve men had died. Magellan himself was killed in a squabble with the inhabitants of the Philippine Islands. Indeed, only eighteen men, traveling on one ship, lived to see Spain again.

However, neither disease nor death stopped the explorers. Success could be profitable, but more than that was the lure of the unknown that kept people going against all odds. The eighteen survivors of Magellan's voyage were under the command of Juan Sebastian del Cano. You might think he had had enough of voyaging, but four years later he was on another voyage into the Pacific, where he died.

Nor was it just a few hardy adventurers that braved it all. Thousands of Europeans crossed the Atlantic, not in a search for gold or for spices or for profits, but just to find new homes where they might live in peace. The payment was heavy. In 1587, a hundred Englishmen, together with twenty-five women and children, settled in Roanoke Island off the coast of North America. Within four years all were gone. We don't know what happened to them—but they were probably killed by Indians.

The English settlement of Jamestown in Virginia in 1607 was more successful. Still, in the ten years between 1607 and 1617, an estimated eleven thousand people came to Virginia, but the population of the colony in 1617 was only one thousand. The other ten thousand had died.

About a hundred Pilgrims landed in Plymouth, Massachusetts, toward the end of 1620, but only a few of them survived the winter.

But nothing stopped the indomitable drive to move outward into the unknown; and, little by little, people learned how to increase knowledge and, with it, safety. Better ships were built. Dietary methods for preventing scurvy were developed. Thus, in the 1770s, Captain James Cook criss-crossed the Pacific Ocean and explored southward into Antarctic waters for three years, losing only one man to scurvy. Still, in 1779, Cook died the same kind of death Magellan did, killed in a squabble with the poeple living in the Hawaiian islands.

In the next century, when the population of the United States surged westward to fill up the new lands that had come under American control, it was at the cost of many lives to starvation and violence.

Meanwhile, the first great explorations of purely scientific interest were being undertaken, as human beings began to explore the Arctic and Antarctic.

At first men poked along the Arctic in an attempt to find another sea route to the Orient, and the first explorations of the Antarctic came because ships were looking for seals and whales to slaughter. This was jettisoned, however, as people became interested in the location of the magnetic poles, in studying the polar environments, and in just the excitement of reaching the poles. (Scientists were also busy exploring the rest of the world, looking for new kinds of plants and animals, to say nothing of other kinds of scientific information. Charles Darwin worked out his theory of evolution as a result of what he learned in the course of a voyage of exploration. He was constantly sea-sick, but he didn't let that stop him.)

Numerous people died in the course of polar explorations. Henry Hudson, who discovered the Hudson River in 1609, died two years later on the shore of Hudson Bay. Vitus Bering, who discovered the Bering Strait and was the first European in Alaska, died on an island in the North Pacific. These are only two of hundreds.

The most tragic story of all involved the two men who were attempting to be the first to reach the South Pole in 1911. One was an Englishman, Robert Falcon Scott, and the other a Norwegian, Roald Amundsen. The story shows what a difference careful preparation makes.

Scott used ponies to help transport his supplies over the Antarctic ice-fields. This was a mistake. Ponies had to eat hay, and this meant dragging great loads of hay. When the hay gave out, the ponies died. And in the last stages of the trek, the men themselves had to pull the sledges.

Amundsen used dogs, which ate the same food the men did. Besides, when food ran low, the weaker dogs could be slaughtered and fed to the stronger ones. As a result, some dogs endured through the entire trip and Amundsen could make much better time. When Scott reached the South Pole, he found that Amundsen had gotten there six weeks earlier. Scott and his men, disappointed and broken-hearted, died in a ferocious blizzard on the way back.

Amundsen got back safely, but he continued polar exploration, dying in 1928 while searching for the survivors of a shipwreck in Arctic waters.

The South Pole wasn't reached by land again till 1958, when Edmund Hillary accomplished it. He had no trouble, however, for he used motorized transport. Technology has removed most of the terrors of polar exploration.

But even now, when humanity has reached a great height of technological achievement, danger cannot be removed altogether, even from devices that are now familiar. Accidents still disable submarines far under the sea, and still destroy airplanes in midflight. Sorrowful headlines mark the deaths of hundreds at sea and thousands in the air and of hundreds of thousands on the highways of the world.

What, then, are we to expect of the new explorations that lie far beyond anything humanity has been able to accomplish in the past? Human beings are now being lifted beyond the atmosphere and brought back, a voyage across a sea of nothingness compared to which the mighty Pacific is a flyspeck.

Even so, they generally do it in safety, and the risks they take are far smaller than those taken by Columbus or Magellan or Scott. The men and women who have been exploring space in the last quarter-century have been in touch with home at almost all times, thanks to radio and telvision. They are not, like the great voyagers of the past, out of touch from the moment they lose sight of land. What's more, the space explorers actually know where they are going, whereas the old ocean voyagers sometimes hadn't the faintest idea of what lay ahead of them. The space explorers need not fear hostile life-forms. Whatever might happen to them, they know they won't be killed as Magellan and Cook were.

But danger has not altogether disappeared. Technology can sometimes fail. Human beings are fallible and make mistakes. Computers may malfunction. Fuel may explode.

And on January 28, 1986, the shuttle *Challenger* exploded and was destroyed, snuffing out the lives of five men and two women as it did so.

The shock was enormous and, in one way, unprecedented. For the first time in the history of exploration, such a catastrophe gave rise to many voices that questioned the wisdom of the human adventure and suggested that humanity draw back.

Why was this?

For one thing, we have been spoiled by success. In twenty-five years of space exploration, no American had died in space. Partly this was due to the use of computers, which have on occasion stopped launches within seconds of zero time because they sensed something was wrong; partly it was due to extreme caution on the part of NASA, with lengthy "countdowns" that tried to take everything into account. This sudden tragedy was worse because it was totally unexpected.

For another, it represented a loss to the public purse. Through the history of exploration, expenses had been borne by private organizations, or, if the state was involved, it was to a very minor amount. Here, a single explosion meant the loss to the taxpayer of a billion-dollar vehicle. Few people could think of that while the tragedy of seven brave people dead filled our hearts and mind, but the thought did eventually creep in.

But the most vital fact was that *this* catastrophe happened before our very eyes. Even those millions who weren't watching when the explosion took place surely saw it in one of the innumerable "replays" that filled the television that afternoon and evening.

No one at home watched Magellan's crew starving, or Scott and his companions freeze, but all of us watched the seven astronauts die.

It was more than we could bear, and yet. . . .

Humanity is more than its members. We are engaged in a great adventure that began four million years ago. We are spreading through the Universe, and we have not yet reached our limits. No disaster has stopped us hitherto, and no disaster must stop us now. The message to the Universe has always been, and must remain:

We are coming!

# 9

# Our Second World

Silence!

The Lunarian stood in the eternal dark within the crater at the Moon's south pole and thought that silence was what was so characteristic—and soothing—and, yes, frightening—about the Moon. He was not yet a true Lunarian, of course. He had come from Earth. And when his ninety-day stint was over, he would return to Earth and try to readjust to its gravity.

There was no motion anywhere, no sound of living things. There was light along the crater top, as perpetual as the dark at this portion of the crater floor. Further along the gently rolling contours of the floor, in the direction of the opposite side of the crater, there was sunlight, too.

The Lunarian looked in that direction, and the glass of his face-plate darkened at once. The line between dark and light swung slowly toward him and away in a four-week cycle. It would never quite reach the point where he was standing, nor ever quite recede out of sight. If he were to move a few miles into the light, he would see the Sun skimming the crater edge along the horizon, but, of course, the face-plate grew virtually opaque if he accidentally looked in its direction. At intervals, he could even see the Earth, or a portion of it, edging above the crater wall. His heart would always melt at that sight.

He tried not to think of Earth. For now, he was on the Moon. He could make out the line of photovoltaic cells in the sunlight and he knew that solar energy, never ending, was powering the world beneath his feet—which was as yet very small. Already, dozens of human beings were housed there, and in his lifetime it might well rise to hundreds. An experimental farm existed there, a chemical laboratory for the study of the lunar soil, a furnace for baking out the small but precious amount of volatiles from appropriate ores.

It was not the only Moon base. A much larger one existed near the lunar equator, where soil was mined and hurled into space for construction. A much more specialized one existed on the Moon's far side, where a huge radio telescope, insulated from Earth's radio interference by two thousand miles of solid Moon, was being completed.

The Lunarian thought: It is 2026 and the Moon has become our second world.

* * *

But it is now 1989. We have visited the Moon six times between 1969 and 1972 and twelve men have trod its surface, but those were visits only. We came, lingered, and left—so that the total time human beings have spent on the Moon is less than two weeks. In the last seventeen years we have spent no time there at all.

But we have been sharpening our space abilities, and when we return, it will be to stay. There will come a crucial day in the future after which there will never be a time, for an indefinite period, when human beings will not be living on the Moon.

NASA is already planning Moon Bases. From April 23 to 27, 1984, for instance, some fifty scientists, engineers, industrialists, and scholars met to discuss scientific, industrial, and sociological issues in connection with such bases.

But why bother? The Moon is a dead, desolate world, without air or water. It is a large super-Sahara, so what is there to attract us and to make us want to go there, let alone live there?

Super-Sahara or not, the Moon would be useful to us, even vital, in many ways. Some of those ways are not material in nature. For instance, there is the question of knowledge. The Moon has not been seriously disturbed during the first half-billion years of the existence of the Solar system (something that is not true of the Earth). We have been studying the eight hundred pounds of Moon rocks astronauts have retrieved—but merely bringing them back to Earth has contaminated them, and the astronauts were able to investigate only the immediate neighborhood of their vessels. If we can study the Moon's substance *on the Moon,* over extended periods of time and over every portion of its surface, we might learn a great deal about the early history of the Moon—and therefore of the Earth as well.

Then, too, the Moon is a marvelous platform for astronomical observations. The absence of an atmosphere would make telescopic visibility far more acute. The far side of the Moon would allow radio telescopes to work without interference from human sources of light and radio waves. The Moon's slow rotation would allow objects in the sky to be followed, without interference from clouds or haze, for two weeks at a time. Neutrinos and gravity waves, together with other exotic manifestations, might be more easily detected and studied there. And, in fact, radio telescopes on the Moon and on Earth could make observations *in combination,* allowing us to study in the finest detail the active centers of galaxies, for instance, including the active center of our own Milky Way.

The Moon can also be used as the site for experiments we would not wish to perform in the midst of Earth's teeming life. Think of the genetic engineering we could perform, of the experimental life-forms we could devise. We could obtain energy in copious quantities for use not only on the Moon but for transfer to space structures and even to the Earth. Think of the nuclear power stations we could build (both fission and, eventually, fusion), where safety considerations did not loom so largely. Think of the efficiency of the solar power stations we could build on a world without an interfering atmosphere to scatter, absorb, and obscure sunlight.

Nor need we think of the Moon as completely desolate. Moon soil, raw and untreated, can be used as shielding against radiation, not only on the Moon itself but in large space structures within which human beings could live and work. The absence of a protective atmosphere in space makes comsic-ray particles an ever present risk, and no form of protection would be cheaper than lunar soil.

Much of lunar soil is in the form of glasses of one sort of another. If these are compacted and reinforced, they could make an excellent construction material. Cement and concrete could also be made out of the soil. Moreover, individual elements can be obtained from the soil. The Moon's crust is forty percent oxygen (in combination with other elements, of course). This can be isolated. For example, a common mineral on the Moon is "ilmenite," or titanium iron oxide. Treatment with hydrogen can cause the oxygen of ilmenite to combine with the hydrogen, forming water, and the water can be broken up into hydrogen and oxygen.

But where would the hydrogen come from? Those portions of the Moon we have studied are lacking in these vital light elements: hydrogen, carbon, and nitrogen. That makes it seem that these "volatiles" will have to be imported from Earth (which has plenty), but there may be places where they can be found in small amounts on the Moon, especially in the polar regions where the Sun rarely, if ever, shines. Lunar hydrogen can then be used to obtain oxygen, and lunar nitrogen can be used to dilute it—and (voila!) there's an atmosphere. (Both hydrogen and nitrogen can be recycled.)

Other elements, particularly iron, aluminum, and titanium—all very useful structurally—are common in the lunar crust and can be smelted out of the soil. In addition, silicon can be obtained for making computer chips. The Moon will be an active mining base to begin with. Quantities of lunar soil can be hurled off the Moon by a "mass-driver" powered by an electromagnetic field based on solar energy. This would not be difficult, because the Moon is relatively small and has a gravitational pull much weaker than that of Earth. It take less than 5 percent as much energy to lift a quantity of matter off the Moon than it would to lift that quantity off the Earth.

Once in space, the lunar soil can be smelted and be used to construct the objects that will make space generally (and not the Moon alone) important to humanity. To build observatories, and factories in space—and, eventually, even artificial space settlements capable of housing thousands of human beings—it would make sense to use lunar materials. These would be far more cheaply obtained than similar material from Earth, especially since earthly resources are badly needed by our planet's own teeming population.

Because of the Moon's feebler gravity, it would be a particularly useful site for the building and launching of space vessels. Since far less power would be required to lift a vessel off the Moon's surface than off the Earth's, less fuel and oxygen would be needed and a much larger percentage of the weight could be given over to payload.

Eventually, when space settlements are constructed, they might be even more efficient as places where space vessels can be built and launched. Indeed, the Moon will have other certain advantages. First, it will be a world of huge spaces and will not have the claustrophobic aura of the space settlements. Second, lunar gravity, though weak, will be constant. On space settlements, a pseudogravitational field based on centrifugal effects may be as intense as Earth's gravitation in places, but this field will complicate matters for settlers by varying considably from place to place within the settlements.

Then, too, since the Moon exists and is already constructed, so to speak, it can surely be developed first and be used to experiment with artificial ecologies. Once the lunar colonists discover how to create a balanced ecology based on a limited number of plant and animal species (and this may take quite a while), that knowledge can be used to make space settlements viable.

Finally, of course, the large Moon, with its enormous supply of materials, may eventually set up a self-supporting system that will leave it completely independent of Earth. Surely, this will become possible sooner than the much smaller space settlements can achieve true independence.

The Moon, as an independent world, will represent a complete new turning in human history. Humanity will have a second world, and that will mean this:

If Earth should be hit by an unexpected catastrophe from without, say by a cometary strike such as that which might possibly have wiped out the dinosaurs sixty-five million years ago; or if humanity's own folly ruins Earth through nuclear war or otherwise; then a second world will exist on which humanity will survive and on which human history, knowledge, and culture, will be remembered and preserved.

* * *

But when will all this take place? Naturally, we can't tell. So much of it depends not on technological ability but on unpredictable economic and political factors.

If all goes well, there is no reason why work on the project can not be initiated in the 1990s. By 2005, the first outpost could be established; and by 2015 a permanently occupied Moon Base could be in existence. After that, it may be that the Moon settlers will have developed their world to the point of being independent of Earth by the end of the twenty-first century.

# 10

# All Aboard for Phobos

The Soviets have sent off two probes in the direction of Mars. It is not Mars itself that they are aiming for, but its two tiny satellites, Phobos and Deimos.

That might seem odd. There is Mars, a world that is 4,220 miles across—twice as wide as the Moon, and half as wide as the Earth. Mars has a surface area of about fifty-six million square miles, which is greater than the land area on Earth. (Of course, we have oceans; Mars doesn't.) What's more, Mars has huge volcanoes, one of which is far larger than anything on Earth. It has an enormous canyon that makes our own Grand Canyon look like a scratch. It has the marks of dried up rivers. It has polar caps made up of ice and frozen carbon dioxide. Everything about Mars is fascinating.

What about Phobos and Deimos in comparison? They are tiny objects, just mountains on the loose. They have irregular shapes that strongly resemble potatoes. The similarity is strengthened by the fact that each satellite is peppered with craters, the larger ones looking exactly like the "eyes" in potatoes.

Phobos is only about sixteen miles across at its longest diameter, while Deimos is only eight miles across. Why, then, should anyone bother with those pipsqueak pebbles when Mars itself is available?

Phobos and Deimos are so small that they weren't even discovered till 1877, long after satellites of distant Jupiter and Saturn were seen. They were discovered by an American astronomer named Asaph Hall, who took advantage of the fact that in 1877 Mars was in part of its orbit that brought it unusually close to Earth (only thirty-seven million miles away). He realized that since no Martian satellites had been discovered up to that point, they must (if they existed at all) be very small and very close to Mars. He searched the neighborhood of Mars for night after night without results, and finally decided to give up, since it was clear there were no satellites.

His wife, however, Angelina Stickney Hall, said, "Try it one more night, Asaph."

You guessed it. The next night was the night. And two of the large craters on these satellites are named Hall and Stickney.

Satirists Jonathan Swift and Voltaire guessed back in the 1700s that, since Earth had one known satellite and Jupiter had four, Mars, the planet in between, must have two. The reasoning was fallacious, but the guess turned out correct—so two other craters are called Swift and Voltaire.

However dramatic the tale of the discovery of those satellites, they are still just mountains on the loose, speeding about Mars. The question remains: Why should anyone be interested in them?

Ah, but there's one thing that Mars *doesn't* have. Two Viking probes sent out by the United States in 1976 showed that there was no organic material, nothing in Martian soil that contained carbon atoms—nothing at all. There can't be life without organic material, and that seems to prove that Mars is a dead planet. Of course, the probes only investigated two small spots on a vast surface, but the chances for life, even anything as simple as bacteria or viruses, seemed small indeed after Viking.

But if there is no life on Mars, how can we possibly expect to find life on Phobos or Deimos? Well, we can't, but there's something else we might find, and to see what that is, let's move back to Earth for a while.

The space between the planets is filled with solid debris of all sorts, a vast number of dust particles and pieces of grit. There are even a great many pebbles, and some boulders, and a very few large hunks of matter that may be up to a couple of miles across.

Some of this material is constantly being swept up by Earth as it moves along its orbit. The pieces of grit heat up as they flash through the air, become white hot and vaporize away. They are the familiar "shooting stars" or "meteors" that can be seen now and then on any dark night if we watch patiently. Sometimes Earth passes through a particularly rich cloud of debris and there is a "meteor shower."

Most of these shooting stars don't survive the trip through the atmosphere, but a few of the larger bits of matter do. These manage to strike the surface of the Earth, and those that do are known as "meteorites."

In ancient times, people were fascinated by meteorites because they seemed to fall from heaven and were therefore viewed as divine. Modern astronomers are fascinated by them because they represent the only samples of extraterrestrial matter they could study before the dawn of the space age.

It turns out that there are several varieties of meteorites. The most common are "stony meteorites," which are not very different, on casual observation, from the rocks of Earth's surface. That means that unless stony meteorites are actually seen to fall, they are not usually discovered unless they land in a region in which the soil is relatively free of stones. (They would easily be found in a place like Kansas, for instance, or in Antarctica.)

Some 8 or 9 percent of meteorites are "iron meteorites." These are easily recognized because lumps of metallic iron are not found naturally (unless they are of human manufacture). In ancient times, before people learned methods for smelting iron out of its ores, iron meteorites were the only possible sources of that metal. Moreover, the meteorites were not pure iron but a much harder nickel-iron mixture. They were far harder and tougher than any other known metal, and finding one was very much like finding a gold mine.

The fact that there are these two kinds make one think that meteorites may be the debris of a shattered planet. After all, planets like Earth, Venus, and Mercury consist of an iron core surrounded by a rocky shell. The iron meteorites may once have been part of a planetary core, and the stony meteorites part of a rocky shell.

If so, might there be meteorites that are part of the actual, planetary surface?

Actually, there is a very rare type of meteorite called a "carbonaceous chondrite." It is black in color, and unlike other meteorites, it is fragile and easily broken up. The most unusual thing about it is that it contains carbon. As much as 2 to 4 percent of its weight may be carbon, which is why it is called "carbonaceous."

Carbon is the characteristic element of life. We can speculate on the possibility of life without carbon, but we have never observed any noncarbon life on Earth, and biochemists would not be surprised if there was no noncarbon life anywhere in the Universe.

We cannot argue the reverse, though. The absence of carbon rules out the possibility of life, but the presence of carbon does not imply that life *must* be present. Carbon may be present but never come together in the supercomplex molecules that are characteristic of life. Nevertheless, if there are meteorites that contain a substantial bit of carbon, scientists want to know what form that carbon takes.

Only about twenty carbonaceous chondrites have been seen to fall and have then been picked up. One important fall came on September 28, 1969, when an object exploded over the town of Murchison, Australia, and showered fragments over an area of fourteen square miles. Eventually about 182 pounds worth of fragments were picked up. In 1950, there had been a smaller fall near Murray, Kentucky.

These fragments, collected before they had had a chance to be contaminated by Earth's atmosphere, were studied carefully, and were found to contain some carbon, fats, and amino acids.

Fats and amino acids are characteristic of living tissue, but their mere existence doesn't indicate that life is present in meteorites or ever was present. Fats and amino acids can be formed by processes that don't involve life, and there are ways of telling from the particular fats and amino acids present whether they were manufactured by life processes or not. It seems clear that the substances in these meteorites were manufactured *without* life.

This is not so terribly surprising, since in the last twenty years, astronomers have located signs of the existence of carbon compounds in vast dust clouds lying between the stars, and there, too, it seems quite certain that the process of formation did not involve life.

But even so, the existence of such compounds is important. If we knew exactly how such compounds came to be formed, we might know what went on in the oceans and air of the very early Earth before life had made its appearance here. We might learn, in other words, how life evolved on Earth.

It is, therefore, very frustrating that there should be so few carbonaceous chondrites to study. It may be, of course, that these objects are not really few, but that there are many of them flitting through space. Unfortunately, they are fragile objects, and if they hit Earth's atmosphere, they tend to break up far more easily than other types of meteorites would. Even those that reach the surface of the Earth in pieces large enough to be studied might have undergone crucial changes in the scorching process of entering Earth's atmosphere. Therefore, even if carbonaceous chondrites are common in space, that would not keep them from being rare here on Earth's surface.

What we *really* want to do, then, is to study carbonaceous chondrites *before* they strike Earth—that is, while they're still in space. It is very likely that meteorites have been moving through space for four and a half billion years without perceptible alteration, and if we could study them in space we would have a marvelous opportunity to learn something about the Solar System in the days of its first formation.

But how can that be done? To send a probe out into space, just looking blindly about for hunks of matter and trying to intercept them and hoping that one of them might be a carbonaceous chondrite is not going to work. Searching for a needle in a haystack would be a simple task in comparison.

Still, it might be possible to search for carbonaceous chondrites from Earth's surface and, *after* we locate them, send out a probe to study them.

Is that possible? Well, between the orbits of Mars and Jupiter is the "asteroid belt," which contains a vast number of small planetary bodies, the largest of which, Ceres, is only six hundred miles across. There may be as many as 100,000 asteroids that are at least one mile across. These asteroids have been studied for nearly two centuries and, little by little, we have been learning more and more about them. For instance, in the mid-1970s it became clear that some asteroids reflected more light than other asteroids do. This is not surprising. Planets with cloudy atmospheres reflect most of the sunlight that hits them. Venus, for instance, with its perpetual cloud layer reflects three-fourths of the sunlight that falls upon it. This reflection, or "albedo," is therefore 0.75.

A body that lacks an atmosphere exposes a bare rocky surface to sunlight, and this tends to absorb most of it. The Moon, for instance, which lacks both air and water, reflects only about one-fourteenth of the sunlight that falls upon it. Its albedo is only about 0.07. (If the Moon had a nice, thick, cloudy atmosphere, it would be ten times as bright as it appears; at the time of the full Moon, Earth would lie in the glow of a kind of twilight.)

Oddly enough, it turned out that some asteroids reflected even less light than the Moon did. Ceres, for instance, the largest asteroid, reflects only one-nineteenth of the light that falls on it, so that it has an albedo of only about 0.053.

Why should the rocks of Ceres's surface reflect less light than the rocks of the Moon's surface? Obviously, Ceres's surface must consist of darker rock than the Moon's surface does. One thing that would make rocks darker would be the presence of carbon. As I've already said, carbonaceous chondrites are black in color. Can it be that Ceres is a huge carbonaceous chondrite, or that its surface, at least, has carbonaceous chondrite characteristics?

Perhaps. Some asteroids reflect light to an even lesser extent than Ceres does. In fact, a few reflect only one-fiftieth of the light that falls on them and have an albedo of only 0.02. Astronomers think that, indeed, many asteroids have carbonaceous chondrite surfaces at the least, and that these are by no means uncommon. Of the twelve largest asteroids, no fewer than eight are suspiciously dark. Of course, these asteroids are quite far away from us. Ceres, for instance, at its closest, is 165 million miles away from us. That's seven hundred times as far away as the Moon is.

We needn't aim for Ceres, however. There might be useful asteroids that are closer to us than Ceres is. And that brings us back to Phobos and Deimos.

Jupiter, Saturn, and Neptune, three of the giant planets of the Solar System, all have one or more satellites that are particularly distant from the planet, particularly small, and have odd or lopsided orbits. Astronomers suspect that in each case, these unusual satellites are really captured asteroids.

It seems unlikely that Mars (a very tiny planet compared to the giants) would capture asteroids. But it is also at the very edge of the asteroid belt and would have more opportunities to do so than one might suspect. Consequently, it is assumed by many astronomers that Phobos and Deimos were originally asteroids that had passed so close to Mars that they were captured.

A point in favor of this captured-asteroid hypothesis lies in the marked difference in appearance between Mars and its satellites. Mars is a bright world with a reddish surface. The asteroids are dark.

It is that difference which points up the fact that, however interesting Mars might be, its satellites mustn't be ignored. Mars is a world of ordinary sand and rocks with a high iron content (that accounts for its reddish—or, more accurately, *rusty*—appearance.) It has no organic material that the Viking probes could detect, which means a very low carbon content. The asteroids, however, being dark, may very well have surfaces that are carbonaceous chondrite in character. They might be rich in carbon and in carbon compounds.

Then, too, Mars has undergone many changes in the course of its history. It must have had a hot interior to have been able to produce volcanoes, and these volcanoes in spewing out molten lava had changed the nearby surface drastically. There are signs of rivers in the past. These are dry now, but there may have been a time when tumbling water changed the landscape. Even today there is a thin atmosphere that is nevertheless capable of creating planet-wide sandstorms.

The two satellites, however, are changeless. They are rocks that have never been exposed to air or water; that have had no internal heat to upset them. They are undoubtedly as old as the Solar System, and very early in the game they were pummeled by other pieces of rock, lit up by the energetic Sun, blasted by a strong solar wind. Those early effects may have produced fascinating organic molecules that have been resting quietly ever since . . . waiting for human discovery.

And what's more, Phobos and Deimos are the closest objects we know of that are, at the same time, at least mountainous in size and very likely rich in carbon. At their closest, they are only thirty-five million miles away, just a little over one-fifth as far as Ceres. What's more, they are eminently reachable, since Mars itself has been reached on several occasions and, from the astronomical standpoint, the satellites are located in the same spot that Mars is.

There is, then, the potential for great excitement in early 1989 when the two Soviet probes reach Phobos. (If the first one is successful in its studies of Phobos, the second will be rerouted to Deimos.) We may find a rich array of organic materials. We may find evidence out of which to construct new notions as to the conversion of inanimate to animate mater. We may find pages in the early history of the Solar System that will prove utterly surprising to us and force us to revise some of our basic notions as to how the planets were formed.

And, of course, we might, on the other hand, be disappointed. The probes may fail us,* or, even if they succeed, they may find that the satellite surfaces are unsurprising and unilluminating.

We must wait and see.

My own feeling, of course, is that even if we do not find organic riches on the surface of the satellites, it is extremely unlikely that we will find only the "unsurprising and unilluminating." Our history of space exploration has, in almost every case, surprised us and introduced us to things we did not expect.

We have discovered the solar wind, the rotations of Mars and Venus, the heat of Venus, the dead volcanoes of Mars and the live volcanoes of Io, the world-wide glacier of Europa, the thick atmosphere of Titan,

*Unfortunately, they did.

the jumbled landscape of little Miranda. Beyond our Solar System, we have discovered quasars, pulsars, and black holes, very distant galaxies, very active galaxies, and even exploding galaxies. Almost every item came as a total surprise.

Why should we not be surprised this time, too?

And what will come next? If those missions to the satellites of Mars succeed, that do we do next?

My own feeling is that, however proficient our instruments are, they cannot match the versatility and intuitive excellence of the human brain. Sooner or later, we must send human beings to Mars.

To be sure, there is danger involved here. A probe that does not carry human beings may miscarry, malfunction, or explode; and we lose only money and effort. With human beings on board, we lose brave lives—the emotional reaction would be extreme, as we found out in the case of the *Challenger* disaster.

Yet even so, we must risk it.

It would be better, at least in my opinion, to wait until we have established a firm base in near space. We would need a working space station, a mining station on the Moon, the construction of various important structures out of lunar material, and (most important of all) the development of a reservoir of human beings accustomed to space and to the idea of space travel.

We can also hope that this expansion of humanity into nearby space will be accompanied by a gradual increase in cooperation between the nations, particularly those nations with the most advanced space technology—the United States and the Soviet Union.

In the first half of the twenty-first century, then, it may be that elaborate human expeditions will be outfitted to visit Mars and perhaps, eventually, to colonize it and to set up scientific stations on its satellites.

This will represent the first extension of the human range beyond the Earth-Moon system and will mark the beginning of something which, by the twenty-second century, will make vacation trips to Mars and forays into the asteroid belt commonplace.

I know this seems fanciful, but think of that we have accomplished in the first thirty years of the Space Age. We began with a single object beep-beeping its way around the Earth in 1957, and now we have a complex probe that has successfully explored Uranus and its satellites and will, within a year, study Neptune, the farthest of the large planets.

Twelve men have already stood on the Moon. Several have remained in space for nearly a year and have survived. And sober plans are being made for projects that, thirty years ago, would have been dismissed as childish, impossible dreams.

Where will we be thirty years from now? Fifty? A hundred?

# 11

# What Do We Do Next in Space?

What do we do next in space? Where do we go? The *Challenger* disaster has sapped the will of the American people and destroyed some of their self-confidence. We hope to return to space, but if so, we have those questions: What do we do next? Where do we go?

One possible answer: We will do nothing and go nowhere. Space exploration costs money, lots of money, and as long as the nations of the world spend half a trillion dollars or so each year on war and preparations of war, there is going to be little money available for space. And if what money is available is spent on converting space into another arena for war, then there will be virtually nothing left for peaceful exploration.

But suppose (just suppose) there is a growing aura of peace and cooperation in the world, that the arms race is ended or at least muted, and space becomes a global project. Money will then be available in suitable quantities, as will time, effort, and resources. *Then* what do we do, and where do we go?

Let us reason it out.

We need a base from which to launch our vessels into space. We have one, Earth, but it is not satisfactory. Its surface gravity is, by definition, one, and the escape velocity is seven miles per second. There are only five bodies in the Solar System with a more intense gravitational field and a higher escape velocity: the Sun, Jupiter, Saturn, Uranus, and Neptune. Though these may be explored by uncrewed vessels, there is no hope in the foreseeable future that human beings will land on these objects or even come very close. Thus, of all the objects we can conceivably use as a base in the foreseeable future, Earth has the deepest "gravity well" and is the most difficult to escape from.

Then, too, Earth has an atmosphere and weather. Storms inhibit launches; even clear air offers resistance. The only object not eliminated by gravity and that has a thicker atmosphere than Earth is Venus (and *possibly* Saturn's satellite, Titan). Venus is so hot that it is unlikely that human beings will ever land on it, and Titan is so far away that we are not likely to reach it for at least a century.

So again, of the worlds we can reach and land upon, Earth is the most difficult base.

What we need is a sizable world, which is, however, smaller than Earth so that it has a lower escape velocity, and which has no atmosphere. As a kindly Fate has it, the very closest astronomical body to ourselves is ideally suited for this. It is the Moon, which has a diameter of 2,160 miles, an escape velocity of but 1.5 miles per second, and no atmosphere. It is only a quarter of a million miles away, and, with present rockets, it can be reached in three days. In fact, human beings have reached it six times so far.

It follows then that, if we are to go into space on a large scale, we must be in a position to use the Moon as a base—it being far more satisfactory than Earth.

But how do we get to the Moon? We have already gotten there, it is true, but that was just the beginning. Going there, staying a few hours, picking up a few rocks, and returning is spectacular—but it is a dead end, as we have, in fact, found out.

We must return to the Moon on a large scale and establish a permanent base there. To do this, we must get there in stages.

We must build a space station in orbit about the Earth, close enough to Earth to be reached easily and frequently, far enough from Earth so that its orbit won't decay and bring it crashing down for up to a million years. If we establish a permanently occupied space station, with shifts of astronauts, we can put together vessels that can then reach the Moon. Such vessels would be launched at lower escape velocities (being farther from Earth) and without atmospheric interference. In this way, we can reach the Moon in two stages and, in the long run, more cheaply and conveniently than if all trips must be made from Earth's surface.

It might seem that if we can build a space station, that would give us our base at once, so we would not need the Moon. Not so. The space station, try as we might, would be small and offer little elbow room. Furthermore, *all* supplies of whatever sort would have to come from Earth where the crunch on resources is already bad enough.

The Moon is a world, with a surface area equal to that of North and South America put together. From its surface material, we can get a large variety of metals, to say nothing of concrete, cement, glass, and oxygen (see p. 41). In fact, elaborate mining stations on our Moon base would supply everything we need for space construction except the light elements, carbon, nitrogen, and hydrogen. These can be supplied from Earth, which must also supply the human beings.

Using the Moon as a base for materials, and the Earth as a reservoir of technology, the space between Earth and Moon can be filled with structures—solar energy stations, nuclear energy stations, observatories, laboratories, factories. We should put as many of Earth's industrial plants into orbit as we can, in order to take advantage of space's unusual properties (vacuum, microgravity, temperature extremes). Such a transfer would hold this extra incentive: The biosphere could be used as a reservoir for wastes. Most of all, space settlements can be built in which Earth's environment can be duplicated as much as possible and each of which my hold up to ten thousand people. (Naturally, there will be problems. Already, for example, "space junk" in the form of innumerable pieces and particles clutter Earth's vicinity and will have to be dealt with, somehow.)

Ideally, this extension of the human range should be global, under international control. The sad history of national strife on Earth must not be duplicated in space. In fact, as the Moon and the space settlements become more populous, international control should be looser, and the new worlds should be regional self-governing units of a Human Federal Union.

It may take a century or more to flesh out the Earth-Moon system into a multi-world home for humanity, but only then will we be ready to take the next major step to Mars. We can, if we wish, launch a Soviet-American manned mission to Mars within a quarter of a century, but it would be a shoe-string affair, difficult to carry through, high-risk in nature, and without important consequences—like the early trips to the Moon.

If, instead, we wait until the Earth-Moon system is solidly established, then the trip to Mars will become relatively simple because it will *not* be carried out by Earthpeople. It was the Europeans who settled the eastern coast of what is now the United States, but it was Americans, the descendents of those settlers, and not Europeans, who later explored and settled the American west. For that matter, it was Americans, not Europeans, who first reached the Moon.

It should not be Earthpeople, then, who go to Mars. It should be the Moonpeople and the Space Settlers. They will be far better suited to the task. They will be much more accustomed to the idea of space flight, since it was through space flight that their homes came into existence. They should be much more accustomed to low gravity and varying gravities. They should be much more accustomed to living on the *inside* of a world, rather than, as on Earth, on the outside. They should be more aware of the necessity of tight recycling of air, food, and water.

In short, when an Earthperson enters a spaceship, he is entering a "world" that is strange and different in every possible respect from Earth. When a Moonperson or a Settlementperson enters a spaceship, it is smaller than home but, in other respects, it is the same. He or she has not had to change viewpoints.

A Moonperson or a Settlementperson is better-fitted physiologically and, far more important, psychologically, for the long trip to Mars. And, having been launched from a low-gravity (possibly micro-gravity) base, there will be less need for fuel, which will leave greater room for control and living quarters.

Mars will be the first world that humanity will reach (other than Earth itself) that will contain substantial quantities of the light elements carbon, nitrogen, and hydrogen. This could make the Moon and Settlements independent of Earth economically, were it not that by that time, it might well have become possible for the people of the Moon and the Settlements to develop techniques that would enable them to capture small comets passing, now and then, not far from the Earth-Moon system. These would be significant light-element reservoirs.

Once well established on Mars, humanity will expand out to the asteroid belt, where a hundred thousand relatively small worlds exist, many of which (most of which? all of which?) can be carved out into settlements or used for further mining operations.

And these asteroid settlements, once equipped with advanced propulsive mechanisms might themselves be steered out into the vast expanses of the outer Solar System or beyond the Solar System altogether. No one making these long trips would be conscious of having left home, for they would be taking home along with them, and humanity would advance to the slow, slow exploration and settlement of the Galaxy, by a process very much like the fluff-supported seeds blown by the wind out of the dandelion.

# 12

# Adventure in Space

The Space Age is thirty years old. In that time we have progressed from placing primitive satellites into orbit to launching complex and sophisticated ones to study Earth and sky, satellites that serve as navigational helps, weather forecasters, and communications relays.

Men have gone into orbit, and six rockets have separately landed human beings on the Moon. Long-distance probes have photographed every planet from Mercury to Uranus, and man-made landers have rested upon the soil of Venus and Mars.

Have we done all we can? Is the adventure over?

Not at all. If we are serious in our efforts to carry human beings, and to expand human knowledge, beyond Earth's surface, and even beyond the Moon, we have an incredibly large Universe beckoning to us.

We cannot remain forever in the Earth-Moon system. Beyond it lies Mars.

## I.

Of all the worlds in the Solar System, Mars is most similar to Earth. Indeed, it would be far easier to explore than the Moon. Mars has a period of rotation similar to Earth's—not a two-week day and a two-week night as the Moon does. It has some atmosphere and water, while the Moon does not. It has a surface gravity only 40 percent as strong as Earth's, but that is still 250 percent as powerful as the Moon's. The atmosphere of Mars provides some protection against meteorites and hard radiation, and it is in any case farther from the Sun and less subject to its radiation than the Moon is.

Disadvantages? Well, yes. Mars, even at its closest, is 140 times as far from us as the Moon is. Where a trip to the Moon by rocket takes three days, one to Mars would take many months. The atmosphere of Mars is not only too thin to breathe; it has no oxygen. What's more, it is a cold world, with temperatures as low as Antarctica's, or lower.

Still, astronauts in spacesuits would be able to roam that kind of world easily. And by building an airtight structure underground, they might remain there for extended periods.

To be sure, the adventure of space is not achieved without cost; *no* adventure is. The *Challenger* disaster has disheartened the United States; and for a year and a half now its space program has been on hold. It may make Americans more cautious in the foreseeable future as far as dramatic projects are concerned.

There is, however, another nation that is interested in dramatic space exploration in a major way, and that is the Soviet Union. They, too, have had occasional fatal failures, but they seem to be more stoic about them.

In any case, they are planning major programs in connection with Mars. They have gigantic launchers equipped to hurl into space objects more massive than either they or the United States have yet handled, and, if all goes on schedule with them, they plan to launch an unmanned spacecraft toward Mars in June of 1988.

That craft will take four hundred days to get to the neighborhood of Mars, but Mars won't be the real target. That will be the little Martian satellite, Phobos (see p. 64). This is a small, cratered world about twenty-five kilometers across at its longest.

Why bother with this little world when Mars is available? That is because Phobos is, in all likelihood, a captured asteroid quite different from Mars in composition. Whereas Mars is a "light" world of rusty rock, Phobos is very dark, probably resembling certain meteorites called "carbonaceous chondrites" that occasionally fall to Earth. That means it may contain water and organic materials, which may give us a hint as to how simple chemicals originated on Earth four billion years ago, chemicals that led to the development of life.

The Soviets plan to pass within fifty meters of little Phobos and to drop a probe to its surface. Then, during the 1990s, the Soviets hope to place "rovers" on Mars that will move about for distances of twenty miles or so at first. More advanced models would later move several hundred miles. In this way, scientists could, within limits, direct the rovers in the direction of objects that seem to hold particular interest. They would no longer have to rely on the "luck of the landing" to determine what they could find and study.

Nor are the Soviets confining themselves to the surface of Mars. They are indulging in dreams of three-dimensional investigations. It might be possible to design light balloons that can be supported even by Mars's thin atmosphere. The Martian winds may drive them thousands of miles, so that they can send back virtually endless panoramas of Martian scenery. There is also the possiblity of landing small burrowing machines that would dig eighteen to thirty meters below the surface and give us an entirely new kind of information about the planet. Toward the end of the 1990s, the Soviets even hope to send probes that would bring samples of Martian rocks back to Earth.

These are only plans, of course, and much can to wrong.* Political and economic developments on Earth may force the Soviet Union to delay the projects—or a launching may fail—or unexpected hitches may develop in some key components of the probes.

Nevertheless, it would be wise, in my opinion, for the United States (and, indeed, for all nations) to cooperate in such large projects. If we do, the projects will be less expensive for any one nation. And whatever portion of the program succeeds can give any of the participating nations a feeling of pride, thus contributing to further cooperation and friendship—which, in this dangerous world, we need desperately.

In fact, on a global basis it might be possible to send a joint *manned* expedition to Mars. That would be the greatest adventure on the immediate horizon.

## II.

Beyond Mars lies Jupiter. Jupiter is the giant of our planetary system. About 70 percent of all planetary mass orbiting the Sun is concentrated into that one planet.

Jupiter is eleven times the size of the Earth in diameter, and it has a mass about 318 times that of the Earth. It is five times as far away from the Sun as Earth is, and it is totally different from the Earth in composition. Instead of being a rocky world with a metallic core, it consists, for the most part, of the two simplest atoms: the gases hydrogen and helium. That is why Jupiter and planets like it are called "gas giants."

*Alas, much has. The first two probes have failed.

A world that is so massive, so far from the Sun, and so different in composition, is bound to be a puzzling object in many ways, and astronomers are eager to find out all they can about it. Jupiter has a cloud-layer marking its visible surface, and it is torn by monster storms. We see the largest of these storms as the "Great Red Spot," a gigantic tornado that has been whirling away for centuries and is so enormous that the entire Earth could plunge into its funnel without touching its sides. Jupiter also has a huge magnetic field that makes the space about it deadly for a million and a half kilometers or more. Making the place even less hospitable is a ring of debris, invisible from Earth, that circles its equator. These are fascinating phenomena, and there's more to be found on each of the planet's four large satellites.

Much of the detail of what we now know about Jupiter was only learned in the last dozen years, thanks to rocket probes. Isn't that enough? What more is there to be learned?

A probe called *Galileo* (which, unfortunately, has been delayed indefinitely because of the *Challenger* disaster) was planned to do much more than earlier probes did. It would supply data from the vicinity of Jupiter not for a few weeks, as the earlier probes did, but for nearly two years. *Galileo* would take advantage of the gravitational pull of the satellites to move into different orbits about Jupiter and, in the process, come closer to the satellites than the earlier probes did. In some cases it would even skim by at fifteen hundred kilometers or less.

*Galileo* would be able to study the world-wide glacier of Europa, the active volcanoes of Io, and the craters of Ganymede and Callisto in far greater detail than the earlier probes did, especially since *Galileo*'s photography system is far superior to those of the earlier probes. It may be able to make out objects only twenty meters across, whereas earlier probes could only see objects five *kilometers* across at the very least.

Naturally, the *Galileo* probe would also study Jupiter itself. In particular it would drop an instrument package through the visible cloud surface, into the atmosphere of Jupiter—the first time any human-made object would ever enter a gas-giant atmosphere. This package could continue dropping for a long time, of course, for the atmosphere is thousands of miles thick. We can't even be sure if there is *any* liquid or solid surface anywhere beneath the clouds, though perhaps we may be able to decide the matter once we get back the information this package sends us.

As the package drops downward, however, the temperature and pressure of the atmosphere will rise steadily. Within a little over an hour of its disappearance through the clouds, it will probably be crushed and melted into uselessness. During that hour or so, however, it would have sent back invaluable information concerning temperatures and pressures of the uppermost part of the Jovian atmosphere—and of its chemical composition, too.

Undoubtedly, if all goes right, we would gain more information about Jupiter and its satellite system in the nearly two years that *Galileo* would be active than in all of history before its time.

As it happens, Jupiter is thought to have about the same composition that the Sun had when it first formed. It may therefore be that the information *Galileo* radios to Earth could enable us to understand more about the Sun, or even about the formation of the Solar System itself.

Then, just as a bonus, it may be that *Galileo,* in passing through the asteroid belt en route to Jupiter, will take a path carrying it near a sizable asteroid named Amphitrite. That will be the first chance astronomers would have to view an asteroid at close range.

### III.

Beyond Jupiter are still more distant planets, and they, too, are being explored. The planetary probe *Voyager 2* has photographed Saturn and Uranus and is heading out toward Neptune, the farthest known large planet, which it will reach in a couple of years. By the time the probe asks Neptune to say "cheese!" it will have been flying through space for more than a decade.

After leaving Neptune *Voyager 2* will continue onward indefinitely, beyond the familiar planets and through the void of interstellar space. It will serve no purpose out there, of course. It will merely be a wanderer.

Nevertheless, astronomers are speculating on the possibility of launching a probe that would be useful even after it had penetrated far beyond the outermost planet. This probe would leave the Earth at a comparatively low speed and contain some twelve and a half tons of frozen xenon (a rare gas found in Earth's atmosphere). This gas would be heated until its atoms break up into electrically charged fragments (ions). The ions would be expelled forcibly, little by little, so that the probe would slowly accelerate in the opposite direction for a period of ten years.

At the end of the ten-year acceleration, the xenon will be all gone, but by that time the probe will be moving at a speed of 360,000 kilometers per hour, or a hundred kilometers per second. It will then be about 9,500 million kilometers from Earth, more than twice as far away as that little, very distant planet, Pluto.

At that point the fuel tanks would be jettisoned and the probe itself, over fifty thousand kilograms in mass, would continue to move outward at a speed that would very slowly decrease because of the distant Sun's weak pull. The probe would continue to drift outward for forty more years until it was nearly 160,000 million kilometers away from the Sun. This is over a thousand times as far away from the Sun as we are.

On board the probe will be a large telescope. Its job: to send us pictures of the stars, taken at increasingly large distances from us, until the final pictures are taken 163,000 million kilometers away. After that, with the probe's energy supply gone, it would continue onward indefinitely, and uselessly.

Of what use will such distant pictures of stars be?

When stars are viewed from different places, the nearer ones seem to shift position compared to the farther ones. This shift is called "parallax." Tbe greater the shift, the nearer the star. By measuring the size of the shift, we can calculate the distance of the star.

Unfortunately, even the nearest stars are so far away that the shift in position is exceedingly small, even when we shift our position as much as possible. We can take pictures of the stars, for instance, when Earth is at a particular point in its orbit and then, again, when it is at the opposite end of its orbit. These two points are separated by three hundred million kilometers.

Such a difference in position enables us to measure the distance of stars up to values of about a hundred light years (900 million million kilometers). These distances serve as a basis for the estimation of the distances of still farther objects by somewhat less reliable methods.

The pictures of the stars that the far-ranging probe would send us will show them at a distance from us that will be up to five hundred times as great as the extreme width of Earth's orbit. By comparing the distant pictures with those we get from Earth, we will see much larger shifts of parallax and will be able to measure accurately the distances of objects as much as fifty thousand light-years away. Our knowledge of the dimensions of the Universe would be enormously sharpened as a result.

Of course, fifty years is a long time to wait, and we can't expect that many of the people present at the launch would still be alive when the final photos arrive. However, humanity in the past has been willing to wait for centuries for the completion of great projects—the great medieval cathedrals, for example. This is a cathedral of another kind and would be worth waiting for.

IV.

Even a probe that reaches out 160,000 million miles from the Sun is only about 1/270th the distance to the *nearest* star. Reaching even the nearest star might well take a lifetime, and it is not likely that individuals would live long enough to survive a round-trip. Such a round-trip might take as long as eighty years, even if we were to use advanced methods of propulsion that we don't have yet.

It may be, however, that the time will come in the near future when "space settlements" will be placed in the Moon's orbit, or even among the asteroids.

Might it not be possible that some of these settlements, making use of advanced methods of propulsion, would "cut loose" and allow themselves to drift away from the Sun indefinitely? They might use controlled fusion as an energy source and pick up additional stores of fuel from comets they might pass.

Any drifting settlement would be a "starship," carrying perhaps ten thousand or more human beings. It would be a little world of its own—and it would not be lonely, for it would not have left home. It would have taken "home" along with it.

Generation after generation might live on such a world as it travelled through space for thousands of years before reaching other worlds circling other suns—worlds they might explore and even settle. Perhaps they might encounter other intelligences, with whom they could exchange information before passing onward in search of suitable worlds without native intelligences.

In this way, human beings would slowly, but inexorably, colonize the whole Galaxy, even the whole Universe if sufficient millions of years are allowed.

Colonizing the Universe! That would be the ultimate adventure!

# 13

# The Distant Flights

I think that our object should be to get Earthpeople out of the space business. We're not really suited to it. We live an abnormal life, stuck to the *outside* of our spaceship Earth. We are spoiled by a constant gravity, and by an ecology so huge that we are not conscious of the cycling of air, food, and water.

The result is that an ordinary spaceship or space station is alien territory to us. The cramped quarters *inside* the hull, the abnormal gravitational pull, and the tight cycling of necessities are all so difficult to live with.

Of course, we can manage a flight to the Moon; the round trip is less than a week. And we can stay in space for at least a year, as long as we're close enough to Earth to remain in touch and to count on rescue in case of emergency.

What about the distant flights, though? Flights to Mars and beyond? We're not really suited to that, and I'm not sure we can make it. What do we do, then?

Here is my suggestion. I would like to see us concentrate on the Earth-Moon system for a while. Let's build space stations in orbit about the Earth, let's build mining stations on the Moon. Let's build observatories, laboratories, even factories, in the space between here and the Moon. Let's even build space settlements in which thousands of human beings can live in Earth-like environments (notwithstanding the pseudogravity induced by rotating a settlement that would produce some un-Earthly effects.)

It may take a century of hard work and concentrated effort, but as we approach the year 2100, we should have a smoothly working space-centered society based on the Earth-Moon system. It will be getting energy from the Sun, and materials (except for carbon, hydrogen, and nitrogen) from the Moon. The settlements will be exporting food and energy to Earth and will supervise the factories that will take advantage of the peculiar properties of space.

What's more, the space settlers will have an environment just suitable for space-flight. They will not only be accustomed to space, but they will be living in what amounts to a large spaceship. They will live on the *inside,* they will be subjected to variable gravity, and they will be conscious of tight cycling of air, food, and water.

They can get into a smaller spaceship and go off on long trips. They will be psychologically suited to it, as Earthpeople are not. They will not experience the radical change in environment in a spaceship that we would.

It is the space settlers, then, that will be the new Phoenicians, Vikings, and Polynesians—the great navigators of the future. It will be they who will reach Mars, the asteroids, the various smaller bodies in the vast reaches of the outer Solar System.

Not us. Not the people of Earth with our abnormal way of life, which ties us down and makes us prisoners.

# 14

# The Telephone in Space

In the Gilbert and Sullivan operetta *Princess Ida,* the three young heroes make fun of the earnest princess, who has established a woman's college. They satirically list the far-out plans being made by the women, and the very first line of their song goes as follows:

> They intend to send a wire to the Moon—to the Moon . . .

Well, of course we can't. Stringing wires to the Moon for either telegraphic or telephonic communication is utterly impractical, for four reasons:

1. A reasonable wire stretching from Earth to Moon would weigh in the neighborhood of a million tons, and the cost and effort of maneuvering such a wire would far outstrip any good it could do.

2. The distance between the Earth and the Moon varies, during the month, from 221,000 to 252,000 miles. If the wire were long enough to reach the Moon at the farthest, there would be a 31,000-mile slack looping down to Earth in the course of the next two weeks.

3. The surfaces of the Moon and Earth are in motion relative to each other. The Earth rotates every twenty-four hours so that the wire would have to make a sliding contact with Earth's surface, and its track would have to circumnavigate the Earth.

4. We *can* communicate with the Moon, and we don't need a wire for the purpose. We use radio waves.

But now let's think of the Moon itself. Suppose the day comes when we have established a series of bases on the Moon and have, in effect, colonized it. We have growing cities that are either domed or, better yet, underground, each with its self-contained atmospheres, water stores, energy sources, food supplies, and so on. How would they communicate?

There would be radio, of course, but suppose we want wire communication for greater privacy, for the carriage of a higher density of messages, for the variety of abilities that would lend us.

In some ways, it would be far easier to wire the Moon than the Earth. The surface of the Moon has an area of about 15,000,000 square miles, which is equal almost to that of North and South America put together. However, it has no rivers or oceans that the wires must span. It is entirely, one hundred percent, land area. Naturally, it has mountains, but with the surface gravity only one-sixth that of Earth's, they are by no means the barriers they would be on our own planet.

Furthermore, on the Moon there are no ice-fields, no storms, no weather in the ordinary sense at all. To be sure, there are extreme ranges of heat and cold, with the surface warming to the boiling point of water at the midpoint of its two-week-long day, and cooling to sub-Antarctic levels by the end of its two-week-long night. These temperatures, however, are surface phenomena only.

We might imagine the wires laid during the night (it is easier to keep warm in the cold than it is to keep cool in the heat), and they would be laid some feet under the surface, where the temperature is always equable. That would protect them from temperature extremes, from the hard radiation from the Sun, and even from the small bits of dust and grit that are always bombarding the Moon's surface in the absence of a protective atmosphere. (Nor would there be any native wildlife to interfere, unwittingly, with the wires; nor any earthquakes, since the Moon is about dead geologically, as well as biologically.)

About the only thing that would damage the wire network would be the impact of a sizable meteorite, and those would be very few.

If we imagine the Moon's telephone system to be thoroughly computerized and to be allied to radio communication where necessary, then each individual upon the Moon could have his own wavelength as well as telephone number. In achieving telephonic contact, you would aim at a person, not at a place. You would hit your target wherever he or she is, provided he or she wishes to receive messages. Furthermore, no one could possibly be lost, since a call for help sent out at a characteristic wavelength would at once inform the world where the lost person was and *who* he was.

All the subsidiary functions of the telephone network could be made full use of. They could be combined with computer outlets to make it possible to get the news, do one's shopping, and take care of all the other things planned for Earth in the telephonic future. Most important of all, libraries could be completely computerized and education could be supplemented in the home, where every human being could obtain information in any field of interest. (See my essay, "The Global Computerized Library.")

All such things have been forecast and will undoubtedly begin to be applied on Earth long before there are Moon settlements. But it will have to advance slowly on Earth, moving from technologically advanced societies to less advanced societies, and within each society from the prosperous to the deprived. The Moon, even though a late starter, will be less densely populated and have a far more homogeneous society, a society which will be far more evenly high-tech to begin with. It is therefore conceivable that the Moon will overtake Earth, as the United States, between 1800 and 1920, overtook Europe. By the year 2100, the Moon might be far in the technological lead as compared to Earth.

The next large piece of real estate that humanity will likely colonize is Mars. It may not be Earthpeople who will reach and colonize Mars in force, by the way, but "Moonpeople." The colonists of the Moon will be more used to spaceflight, more accustomed to low gravity. In short, they will be, as I said, more high-tech, and therefore readier to undertake the long flights to Mars.

Mars is in some ways preferable to the Moon. It is even larger, with a surface area three times that of the Moon and about equal to the land area of Earth. While it has no rivers or oceans, it does have small polar ice-caps, but they can easily be avoided. It has a thin (unbreathable) atmosphere and sandstorms, but it's farther from the Sun and does not suffer from heat. In fact, it is mostly Antarctic in weather, but again, that is a thing of the surface only. Its surface gravity is 2.5 times that of the Moon, but is still only two-fifths that of the Earth.

Because of its distance from the Earth-Moon system, Mars will undoubtedly lag behind the Moon for a long time in terms of its technology. However, its greater size, its greater supply of elements such as carbon and nitrogen, its lesser exposure to heat may, in the long run, make it humanity's dominant high-tech world.

In addition to the three large worlds—Earth, Moon, and Mars—there are sure to be subsidiary human-occupied objects, especially in the Earth-Moon system. There will be space-stations circling Earth in fairly close orbit, and these will serve as platforms from which the Moon base will be established. These platforms can serve as laboratories and observatories, taking advantage of the unique properties of space. Or they could be solar power stations, automated greenhouses for the growth of plants, factories of all sorts, and so on.

There may even be large space settlements with artificial environments much like that of Earth, each capable of holding ten thousand human beings. These are likely to be in the lunar orbit either 60 degrees ahead of the Moon or 60 degrees behind. These are the so-called "L4" and "L5" positions and are gravitationally stable.

The question arises, though, as to communication *between* the worlds. In principle, the problem is the same as that of ship-to-shore communication, a combination of wire and radio waves. We have seen it work, after a fashion, already, when we were able to see the astronauts on the Moon from our homes—"live."

There is one problem, however, that is undefeatable. Communication *of any sort* cannot take place faster than the speed of light. Since the speed of light is 186,282 miles per second, that does not trouble us on Earth. Electric currents and radio waves can cover the distance from New York to Tokyo in about one-eighteenth of a second, so we can talk easily between those two cities.

However, it takes radio waves 1.28 seconds to travel from the Earth to the Moon, and another 1.28 seconds for them to travel back. This means that if you are speaking to someone on the Moon, you will say "Hello" and then have to wait two and a half seconds to hear a responding "Hello." *Nothing can shorten that interval.* It is something people will, perforce, have to get used to.

There will be the same interval if a conversation is proceeding between people in the L4 position and either the Moon *or* the Earth. If someone in the L4 position wants to speak to someone in the L5 position (they are 412,000 miles apart), then the round trip for communication is 4.4 seconds—and nothing can shorten *that* interval either.

The real problem will arise in connection with Mars. When Mars and Earth are on the same side of the Sun, they can sometimes be as close together as 35,000,000 miles. Mars is then about 150 times as far from the Earth as the Moon is. It would take communication *of any kind* about three and an eighth minutes to go from Earth to Mars, and an equal time to come back. Once you've spoken to a Martian colonist, you will have to wait six and a quarter minutes for an answer, and nothing can shorten that interval.

And that's Earth and Mars at the closest. When Earth and Mars are on opposite sides of the Sun, they could be as far as 250,000,000 miles apart, and the time between statement and answer will be forty-five minutes. In fact, since the Sun will be in between the two planets and may interfere, communication may have to be via satellites circling the Sun at right angles to the orbits of Earth and Mars. That will make the path longer and delay the answer further. You will certainly have to wait over an hour to get an answer.

The situation gets rapidly worse, in this respect, as we penetrate the Solar System beyond Mars, and no conceivable advance in technology will help us. Sorry.

# 15

# The Average Person as Astronaut

Travel in space? For anyone with a ticket?

It still sounds uncomfortably unlikely. It's thirty-two years since human beings first lobbed an object into orbit around the Earth, twenty years since human beings first stood on the Moon. But the average person seems no closer than ever to being an astronaut on demand.

Then, too, even if we *could* travel in space at will—just buy a ticket at the spaceport and take off—where would we go?

The only nearby object in space is the Moon. It takes only three days to get there. But—frankly—once you've seen one crater and one moonscape, you've seen them all. It's just airless, waterless desolation.

Anywhere else? Mercury is like the Moon, except a lot worse, considering the heat and radiation of the nearby Sun. And Venus, although farther from the Sun than Mercury is, is even hotter and much worse: Its carbon dioxide atmosphere is ninety times as thick as ours, its temperature is hot enough to melt lead, and its clouds are rich in sulfuric acid. Dante's Inferno is Palm Springs in comparison.

In the other direction, Mars is better than the Moon, though not exactly livable, but it is some nine months away under present circumstances. We may work out ways of geting there a little faster, but it's hard to see how Mars could be worth the journey for the casual visitor. As for anything farther than Mars—that would mean a journey that would last for years.

We might catch an occasional comet or asteroid on the fly, as it zooms past the Earth-Moon system, but, honestly, there wouldn't be much to see.

Well, then, shall we just leave everything to the professional astronauts and to unmanned probes?

Certainly not! Space isn't going to stay the way it is, and the reason for that is the Shuttle.

* * *

The Shuttle is a reuseable spacecraft. It is a workhorse, a truck for the vacuum. It can carry human beings and material into orbit and come back for more. With enough Shuttles working away, we can eventually carry enough material out into space to put together, let us say, a solar-station designed to collect solar energy, convert it into microwaves, and beam them down to Earth for conversion into electricity.

It will mean a huge initial investment, especially since we will need dozens of such stations, but the energy we collect will quickly make the complex first self-supporting and then enormously profitable. We will not even have to draw upon the Earth itself for the metals and other materials needed to build these stations. Thanks to the Shuttle and to its improved and more sophisticated successors, we will be able to set up a mining station on the Moon.

Material from the Moon can be launched into space with "mass-drivers" (not science fiction, but devices making use of present-day electromagnetic know-how). In space, the Moon-material can be converted into soil, concrete, glass, and a variety of metals.

Observatories and laboratories will also be built in space to study the Universe and to carry through experiments that will make use of the special properties of space—high and low temperatures, hard radiation, vacuum, zero-gravity.

For that matter, automated factories can be built in space in such a way that most of Earth's messier industrial processes can be lifted from the planetary surface they pollute and put in space—where they will offer us the benefits of industrialization without the dangers.

(In view of what happened to Skylab, it might be well to emphasize that all of these will be in orbits high enough to ensure their remaining in space for many millions of years.)

In order to build and maintain all these space structures it will certainly prove desirable to build space colonies—self-contained worlds, perhaps as large as several miles in circumference and capable of holding ten thousand to ten million people.

These settlements will most likely be in the Moon's orbit, either a quarter-million miles ahead of it in its movement around Earth, or a quarter-million miles behind it, the two most stable positions.

If we look ahead to 2080, then, we won't be seeing near-space as it is now—the Earth, an empty Moon, and no more. We will instead see an occupied and busy Moon, with dozens of space settlements preceding and following it, along with hundreds of power stations, observatories, laboratories, and factories, all in orbit between the Earth and the Moon.

* * *

Will all this really happen?

Yes, it will, unless human beings deliberately decide *not* to do it, either out of short-sightedness or a failure of nerve.

There will be those who will want to "solve Earth's problems first." But those problems are insoluble unless we move out into space. We are overcrowding a depleted Earth, which can no longer support us. Unless we expand our range and draw on resources beyond Earth we will collapse, and civilization will wither and die.

But even if we try to do it, can we do it in only one century?

Without question! Imagine yourself back in 1869, when the biggest technological marvel was the Atlantic cable. The electric light was still ten years in the future, the automobile twenty, and the airplane thirty-five. Yet counting from 1869, one hundred years, a single century, saw a man standing on the Moon—and we are moving faster these days.

* * *

If, then, the world of 2080 is as I visualize it to be, people will certainly be able to engage in astronautics at will (provided always they can get their spaceship reservations and afford a ticket).

With solar energy from space, we will have ample electricity to split water into hydrogen and oxygen. If the hydrogen is used as fuel it will recombine with oxygen to form water again. Thus our spaceships will be run, and all we'll use up is sunshine, and that will last for billions of years.

And where do we go?

Once we've exploited and colonized space, there will be no shortage of destinations. The Moon will no longer be a mere desolation. To be sure, it will still be airless and waterless. The time may come when it will be "terra-formed" and made into an Earthlike, livable world—but probably not by 2080.

Yet, even though airless and waterless, it will be a busy mining station. Tourists, in large enclosed ground vehicles, can be carried over the lunar surface to see the mines, the automated machinery in action, the mass-drivers.

There will undoubtedly be elaborate hotels under the lunar surface, where tourists can be completely comfortable in a thoroughly simulated Earth-environment, even to Earth-scenes projected on "windows"—but there will be a difference in gravitation. The Moon's surface gravity is only one-sixth that of Earth's and it will be a long, long time before anything can be done about that—perhaps nothing can ever be done.

But low gravity is not necessarily bad, especially if you are a visitor and are subjected to it only temporarily. There will be exercise rooms in which gymnastics can take on a new and weird beauty, in which people can swing more slowly through more graceful arcs, in which trapeze artists can make more turns, in which ballet dancers can indulge in more spectacular leaps.

The whole world would become slow-motion, in a sense.

Yet the visitor cannot, and dare not, try too much too soon. One must learn how to handle objects and one's own body under low gravity. Weight decreases but mass does not, and the illusion of featheriness will not keep you from breaking your ankle if you land wrong.

But there will be instruction sessions, of course, and they will be eagerly attended, for there will be nothing like them on Earth. The simple act of ballroom dancing can take on gymnastic overtones that simply cannot be duplicated on Earth. Although a low-gravity environment might seem a temptation to indulge in muscle underuse, the fun of it all might actually encourage people to shed weight and firm up their muscles.

The time may come when "Get yourself in shape on the Moon" will be the slogan of the beautiful people of the twenty-first century.

* * *

Nor will the Moon be the only destination. There will undoubtedly be a steady stream of visitors to the automated factories (or laboratories, or observatories) of the future, where automated guide-systems will lead people through the airless caverns where all the silent operations will take place—from the arrival of raw materials from the Moon, or finished components of great sophistication from the space settlements (for machinery must be maintained and repaired), to the loading of products on large shuttles for delivery to Earth, all without the touch of human hands, but under intense human supervision by closed-circuit three-dimensional television.

Undoubtedly, however, the *pièce de resistance* of all space adventures for traveling Earthpeople will be the space settlements themselves.

By 2080, there very likely will be dozens of them both preceding and trailing the Moon, and they will house the busy spacepeople of the future. The space settlers will be controlling the Moon-mining operations and will run the tourist facilities on the satellite. It is they who will construct and automate the factories, laboratories, and observatories in orbit. It will be they who will build new space settlements.

Each different space settlement will be like a different island in the ocean. Each will have its own ways, its own culture, its own "feel."

A space settlement can be designed to mimic Earth inside its enclosing metal in any of a myriad different ways to fit the tastes of the initial settlers. One might be a bit of American suburbia, another might be a Dutch landscape, or an African, or Spanish—different languages, different customs, different attitudes, different amusements.

It will be to the delight of humanity that these small and isolated worlds will add to the diversity of human culture, and therefore to the interest and excitement and adventure of travelers.

No doubt different Earthpeople will have their favorite settlements, but there will always be interest in trying a new cuisine or a new musical beat or a new vista (for even if a settlement is only a few miles across there will be room for vistas).

Space settlements will be rotating, so the centrifugal effect on the inner surface will lend the feeling of Earthly gravity. Mirrored sunlight through louvred windows will be adjusted to give an Earthly succession of day and night. The landscape will be Earthlike in one variety or another. Nevertheless, things won't be *too* Earthlike.

No matter what the geometry of the space settlement—whether cylindrical, spherical, toroidal, or whatever—there will always be places where the pseudogravitational pull will be less than Earth's.

It won't be the obligatory one-sixth-normal of the Moon's surface. It can be anywhere, if you choose your position carefully, from 1 (Earth-normal) to 0. That means there will be places in any space settlement where you can outdo the gymnastics possible on the Moon.

It is quite possible that some settlements will deliberately be designed to offer large surface areas where mountains (full-size) are constructed for the pleasure of climbing. Here, some of the discomforts are gone. In the enclosed space settlement, air will be present everywhere at nearly constant density, so that as you climb your mountain, you will not be subjected to the hazards of thin air or, for that matter, of low temperature. Furthermore, if the mountain is properly placed, the higher you go, the lower the gravitational pull and the easier the further climb.

On the other hand, a fall could carry you downward into regions of higher gravity, so that death might be the consequence of carelessness or accident. (Presumably, the true mountain-climber would not want all the spice of danger removed.)

There might also be large surface areas that are reasonably level but where the gravitational effect is virtually zero. Here it would be possible to use tough lightweight plastic membranes stretched over a lightweight skeleton of flexible plastic rods as "wings" so that you could fly. It would not be the flying that comes of sitting in a machine; it would be personal flying by the use of arm muscles for propulsion and leg muscles for steering.

It won't be as easy as it sounds, for manipulating and handling the wings would be at least as hard as learning to ride a bicycle—but perhaps no harder, if you're interested in no more than belly-flopping through the air. To become an expert, to soar and dip and glide and turn with effortless ease would, like anything else, require considerable practice.

Space travel itself would take on new dimensions once you were on a settlement. On Earth, there is the problem of breaking away from a strong gravitational field. This requires a powerful machine and an uncomfortable acceleration.

The space settlements themselves are firmly in the grip of Earth's gravity, but because they are in orbit, and therefore in "free fall," they don't feel it. The gravitational pull of the settlements themselves is virtually zero, however, and if you are traveling from settlement to settlement and remain at roughly the same distance from Earth you will require virtually no power.

The push away from a space settlement will involve no more effect than that of pushing a rowboat away from a dock; nor will stopping at another space settlement require more. In between will be a period of the greatest possible calm, surrounded by all the illimitable vastness of space. Space settlers are likely to be confirmed space-yachtsmen, and people from Earth will surely love the adventure.

There will be danger here, too. There will have to be weather reports. No, there won't be big winds and rising seas—but there is the Sun, and it is the source of an outpouring of charged subatomic particles in every direction: the solar wind. Ordinarily, the solar wind is not dangerous, but the Sun has its storms. Occasionally "flares" will appear on its surface, and these will blast out floods of speeding particles energetic enough, sometimes, to reach the cosmic ray range. If these floods are in Earth's direction, the normal solar wind will become a gust or a temporary gale. You won't feel it, you won't know it's there, but it could fatally damage your body.

A light space-yacht is not likely to be sufficiently shielded against flares, but by 2080 we will know enough about the Sun to be able to predict its behavior far better than we can now. When a storm is projected in the solar wind, the space-yachts will disappear from space till it passes.

* * *

And, of course, there will come a time, not long after 2080, when the space settlements will feel themselves to be overcrowded and when plans will be made to reach and colonize the vastly more roomy (and safer, because more distant from the Sun) expanses of the asteroid belt.

But that is beyond the coming century and beyond the scope of this article.

# 16

# Other Intelligent Life?

People throughout history have usually assumed that if other worlds existed, every one of them carried intelligent life—usually life that is quite human in appearance.

In the second century A.D., for instance, the Syrian writer Lucian of Samosata wrote the first interplanetary romance that we know of. He tells of a ship that was carried up to the Moon by a waterspout. Was the Moon inhabited by humanlike intelligent beings? You bet it was, and they were at war, too. Their enemies were the intelligent beings of the Sun, and the occasion of the quarrel rested with conflicting ambitions over the colonization of Venus.

It wasn't until the 1600s, and the coming of the telescope, that astronomers were able to show, pretty definitely, that the Moon lacked air and water and could not, therefore, be host to our kind of life. It was the very first time the concept of a "dead world" had been advanced.

Further astronomical discovery finally demonstrated beyond all doubt that the Earth was the only world in the Solar System that bore any life at all, however simple, let alone *intelligent* life.

* * *

But what about the vast spaces outside our Solar System? After all, the Sun is only one star, and there are incredible numbers of other stars. Might there not be life, even intelligent life, associated with some of them? We have no evidence one way or the other, but perhaps we can work out the chances. Let's try.

In the first place, we can ask how many stars there are in the Universe. Then we can have some idea of how many possible places there are where intelligent life may have developed.

Stars are gathered together into vast groups, called "galaxies." Our Sun is part of the Milky Way galaxy, but there are others. Some are vast assemblages of a trillion (1,000,000,000,000) stars, and some are "dwarfs" of a mere five billion (5,000,000,000) stars. The Milky Way galaxy is somewhere between. It and the nearby Magellanic Clouds (its small satellite galaxies) contain a total of 140 billion (140,000,000,000) stars.

We don't know how many galaxies there are altogether. Our best telescopes can detect hundreds of millions, but there must be many others we cannot yet detect. Some astronomers estimate there may be as many as a hundred billion (100,000,000,000) galaxies in the Universe.

If that is so, and if our own galaxy is average in size, then the total number of stars must be something like 14 billion trillion (14,000,000,000,000,000,000,000).

That, however, is a very chancy figure, since we don't *really* know the total number of galaxies. Besides, the other galaxies are anywhere from millions to billions of light-years away, whereas the stars in our own galaxy are within 150,000 light-years of ourselves. It is much more likely that if there are other intelligences out there, it will be those of our own galaxy that will interest us rather than those of the distant galaxies beyond.

Let us, therefore, calculate how many intelligences there are in our own galaxy and, if we do that, we can assume that (on the average) that's how many there are in every galaxy. So we begin with our first figure:

*1. The number of stars in the Galaxy = 140,000,000,000*

A star is essential to the development of life: It is the source of the energy that brings about the development of life, that maintains it throughout its existence. Not all stars are ideally serviceable in this respect, however.

Most of the stars in any galaxy are relatively small objects, dim and red, the so-called "red dwarfs." For life to receive enough energy from such a star, the world upon which the life exists would have to hug the star in a close orbit. The world would then receive enough energy, but it would also be subjected to tidal effects that would slow its rotation and cause it, finally, to face one side perpetually to the star. One side would then be too hot for life; and one side would be too cold.

There are a number of stars that are larger and hotter than the Sun, but the larger and hotter a star is, the sooner it will explode and collapse, and the shorter the total time it will remain a stable object delivering warmth in the steady manner that life requires. If our experience of life on Earth is at all typical (and unless we assume it is, we cannot make any calculations at all), it takes a long time to evolve intelligence. Large, hot stars don't offer enough time. Worlds circling such stars may conceivably develop a thin scum of primitive life in the oceans, but that's not what we're after.

What we want then are Sunlike stars, stars that are not less than a quarter as massive as the Sun, or more than one and a half times as massive. Such Sunlike stars are not very common, but fortunately they are not very rare either. We can reasonably estimate that 10 percent of the stars in our galaxy are Sunlike; so, there's our second figure:

*2. The number of Sunlike stars in the Galaxy = 14,000,000,000*

The nature of a star is not all that counts, however. There is also the matter of its location.

Until recently that might not have seemed to matter, but in the last twenty years the new science of radio astronomy has taught us that the Universe is a much more violent place than we imagined—and that is particularly so in the galactic nuclei, where the stars are packed tightly together.

Galactic nuclei seem to be given to explosions that pour inconceivable energies into space all around them. The mysterious quasars, which burn with the light of a hundred galaxies, seem to be galactic nuclei so violently brilliant that they can be seen at distances of billions of light-years. There is a strong suspicion that at the center of galaxies are black holes that are steadily consuming matter and, in the process, liberating floods of x-rays.

There is a hot spot in the center of our own galaxy, for instance, which could be a black hole as massive as a hundred million stars.

If this is so, then the central regions of a galaxy may be no place for so fragile a phenomenon as life; there is too much radiation there. It could be that only on the outskirts of a galaxy—in the spiral arms of our own galaxy, for instance, where our Sun is located—is it quiet enough for life to sprout and grow.

The central regions of a galaxy contain most of the stars. It is estimated that 90 percent of all the stars in our galaxy are in the nucleus and only 10 percent in the spiral arms.

If the Sunlike stars are distributed in this fashion (and why should they not be?) then only 10 percent of them are in the life-bearing region, and we have our third figure.

*3. The number of Sunlike stars in Galactic outskirts = 1,400,000,000*

A star, however suitable in size and location, does not itself bear life, of course. There must be a planet circling it that bears the life. How sure can we be that any particular star will possess a planetary system?

For the first forty years of the twentieth century, the main thrust of astronomical thought was, in fact, that planetary systems were rare. It was argued that the natural collapse of an original collection of gas and dust through the force of its own gravitation would merely form a star. For planets to form, the star must later be subjected to some catastrophic event, such as a near-collision with another star, that would tear material out of the central body and form the planets.

Such catastrophic events would, however, have so tiny a chance of taking place that even a single such event in the lifetime of a particular galaxy was doubtful. Under those circumstances, it was rather fashionable to think that our Sun might be the only star in the Galaxy to possess a train of planets (plus, perhaps, the unknown star that nearly collided with the Sun, for it might also have given birth to planets in that violent coition).

Beginning in 1944, however, astronomers began to change their minds. The manner in which clouds of dust and gas would condense into a star and the role played by magnetic forces made it seem that there would be turbulence instead of a smooth collapse. The turbulence would automatically form planets on the outskirts even as the star was forming in the center.

If this newer view is so, virtually every star would have a planetary system. Can we decide between these two views?

We just might. When a planet moves around a star, the star jiggles slightly in response. The smaller the star and the larger the planet, the greater the jiggle. And if the star is close enough, that jiggle might just be large enough to detect. In the last quarter-century, some half a dozen small stars that are relatively close to us have shown these jiggles.

To discover so many cases among the nearby stars makes it seem very likely that planetary systems are extremely common and that the newer theory is correct.

Yet a condensing cloud of dust and gas may sometimes form *two* stars rather than one. In facts half the stars in the Galaxy may be part of such a "binary." (A binary may be further associated with a single star or another binary at a great distance.)

If the separate stars in a binary are widely separated, each may develop a planetary system undisturbed by the other. If, on the other hand, the stars in a binary are close together, as many of them are, no stable planetary orbits of a kind that would make for life could be possible. Such stars might have planets, perhaps, but not of a kind that would interest us.

Let us suppose, then, that half the binaries, or one-quarter of all the stars in the Galactic outskirts that are Sunlike, do not form planetary systems of the kind in which we are interested. That still leaves about three-quarters that are potentially suitable. That gives us our fourth figure:

*4. The number of suitable planetary systems in the Galaxy = 1,000,000,000*

Given a planetary system, how likely is it that at least one of the planets it contains will be suitable for the development of life?

Not all planets are suitable. In our own planetary system, only Earth is. Venus, which is virtually Earth's twin in size and composition, is a little closer to the Sun and has grown too hot. Mars, which is very like the Earth in some ways, is a little farther from the Sun and a little smaller, and these facts eliminate it. The Moon, which is at just the distance from the Sun we are, is distinctly smaller and therefore dead.

In short, what we need is an Earthlike planet in size, composition and temperature. In addition, its orbit must not be too elliptical, or its rotation too slow, or its axis too tipped, since any of these characteristics will make weather changes too extreme for life, even if the average temperature is suitable.

There is no way of telling how likely it is that all these rather tight conditions will be met in any given planetary system. We know only our own in detail, after all, and it may not be a fair sample. It is conceivable that the proper conditions are so difficult to meet that only the most fortunate concatenation of circumstances has produced the one Earth we know and that it is the only truly Earthlike planet in the Universe.

That, however, seems an unlikely depth of pessimism. We might strike some reasonable balance as a matter of intuition. We might guess that perhaps one planetary system out of ten will include an Earthlike planet suitable for life, and that gives us our fifth figure:

*5. The number of suitable Earthlike planets in the Galaxy = 100,000,000*

Given an Earthlike planet suitable for life, how likely is it that life will, in actual fact, develop?

Life itself might seem so miraculous a development that it is easy to suppose that it came about only through some supernatural creation; or that, if it comes into being by chance; it is such a vastly improbable chance that Earth may be the only life-bearing planet out of the hundred million potential life-bearers in the Galaxy.

Beginning in the 1950s, however, the scientific view of the probability of life forming underwent a drastic revision. Small-scale experiments in the laboratory have been conducted with a mixture of simple compounds of the type that must have been present on the primordial Earth before life began. This mixture was subjected to the kind of energy one would expect from the primordial Sun or from the Earth's volcanic heat, or from its lightning, or from its radioactivity.

The result was a fairly rapid combination of the simple compounds into somewhat more complex ones. Beginning with these somewhat more complex ones and repeating the experiment leads to the production of still more complex ones.

The most complex compounds formed in the laboratory in this fashion are very far from representing even the most simple conceivable life-form, but they point in the right direction. And if so much can be done on a small scale in the laboratory in a period of weeks, think what could happen in an ocean of materials over millions of years.

In the 1970s, rather complex compounds, undoubtedly formed from simple precursors by processes having nothing to do with life, have been found in meteorites and even in the vast dust-clouds lying between the stars—and these, too, point in the direction of life.

There are strong indications, now, that life is a very natural consequence if you start with a mix of common chemicals in an Earthlike environment. Indeed, there are signs in the rocks that life began on Earth only a few hundred million years after it achieved its present form. If we consider the total length of time during which Earth may have been capable of supporting life, say ten billion years, life developed when that lifetime was only one-twentieth of its way along.

It would seem that 95 percent of the suitable Earthlike planets in the Galaxy would then be old enough to have developed life and that gives us our sixth figure:

*6. The number of life-bearing planets in the Galaxy = 95,000,000*

It would seem that this is a tremendous number, but actually we are not giving ourselves much in the way of breaks. Our line of argument makes it appear that only 1 star out of every 1500 in the Galaxy shines down upon some form of life.

But then, granted the existence of life-bearing planets, on how many of them does an *intelligent* species evolve? Specifically, on how many of them does a technological civilization arise that is capable of interstellar travel?

Again, we have only our own world to go by, and we cannot tell how typical it may be. The development of intelligence could be so chancy a thing that, even though the Galaxy is rich in life, we may be the only intelligence. But such pessimism is, again, unwarranted, and it makes more sense to suppose that the example of Earth *is* typical.

Our Sun has now been in its present Sunlike stage for about five billion years and will continue for five billion more, we think, before it consumes enough of its fuel to expand to the red giant stage and make the Earth no longer habitable. It is for this reason I said, a few paragraphs above, that Earth's total history as a life-bearing planet would be ten billion years.

It took, then, half of Earth's life-bearing history for it to develop a technological civilization. To be sure, we can't manage interstellar travel, but we may be able to in a few centuries. Even if it took us a few thousand years to do it, or a million, that would be a very small part of Earth's total life-bearing history, we could still say that a technological civilization capable of interstellar travel rose halfway into the history of our planet.

If this is an average accomplishment, and we have no real choice but to assume it is, then, on the average, half the Sunlike stars in the Galaxy have reached the halfway point or beyond in their lifetime, and half of them must have given rise to civilizations more advanced than our own. That gives us our seventh figure:

*7. The number of suitable extraterrestial civilizations arising in our galaxy = 47,500,000*

The thought of all those civilizations, all capable of interstellar travel, introduces some complicating thoughts. Does each one spread out on colonizing missions? What happens when they meet? Do they struggle for mastery, and does the victor wipe out the vanquished? Do the civilizations that form first preempt suitable planets and actually prevent other civilizations from arising?

Well, let's go on.

Suppose a technological civilization is established. How long will it survive?

Again, we have only our own case to study. We are in the very earliest stages of a technological civilization. We have only reached one other world, our own Moon, and interstellar travel is as yet out of the question—yet we can already see a good chance that within half a century, our civilization will break down.

The breakdown may come through nuclear war, through pollution, or through overpopulation. In any case, it would mean that we will never have interstellar travel.

Is this a general rule? Is intelligence a suicidal blind alley in evolution? Does every technological civilization break down as soon as it develops?

Surely this is another example of unwarranted pessimism. With nearly fifty million technological civilizations making a beginning, is it unlikely to suppose that one out of a million will reach the interstellar-travel level without breaking-down? If so, might they not continue to exist for as long as their planet remains habitable and, in fact, be in existence right now. That gives us our eighth and final figure:

*8. The number of advanced long-lived extraterrestrial civilizations in our galaxy = 50*

These fifty civilizations will be a special breed, surely. To have survived will have meant overcoming those impulses within themselves that might have led to destruction. They would be humane civilizations, not given to violence, and respecters of life.

I imagine they might colonize suitable planets peaceably, mark off spheres of influence among themselves amicably, avoid planets with developing intelligences of their own, and may even have formed a Galactic League of Civilizations.

And even though there are only fifty, or, for that matter, only a couple, they could, if they have been in existence long enough, have filled the suitable planets of the Galaxy. In other words while the number of existing civilizations may be few, there might be many planets bearing civilizations.

* * *

How can we decide whether all that I have worked out in this essay is actually so?

We can't very well go out there and look, for we don't have interstellar travel, and we are not likely to have it for a considerable time to come. We can't even send out a signal, for we have not yet reached that stage of technological development where we can afford the energy required to send out one that is sufficiently strong.

We might simply wait, of course. If there are indeed civilizations out there, surely some extraterrestrial ship is bound to come to us eventually.

The fact that none has yet done so would indicate that a) there are no civilizations out there, after all, or b) there are, but the Galaxy is huge and they haven't found us yet, or c) they *have* found us, but they are leaving us to ourselves to see if we overcome our difficulties, develop interstellar travel, and earn a worthy place for ourselves in the Galactic League.

(There are some people who think that extraterrestrial ships are reaching us now, as flying saucers, or that they have already reached us in prehistoric times—but the evidence for this, to anyone but a blind enthusiast, is totally unconvincing.)

Since we have no idea when we will be found, or when we will be judged worthy, there remains only one thing to do if we cannot bear to wait, and that is to search for *their* signals. There is almost no likelihood that they will be signalling to us directly, but they might be signalling casually, or even inadvertently, as part of their ordinary lives. If we pick up, let us say, microwaves that are neither absolutely regular nor absolutely random, they will then contain "information," and we will have a strong indication that an intelligence is behind it.

Ever since 1960, astronomers have occasionally scanned the sky in the direction of certain Sunlike stars in the search for signals of intelligent origin and have found none. NASA now proposes to spend five years and $20,000,000 to listen to every Sunlike star within a few hundred light-years of ourselves.

We *may* hear something.

# Part III
# Science

I have been writing science essays for nearly forty years. It began this way:

By 1953, I had already decided that I had no real aptitude for scientific research, yet I needed scientific papers if I was to maintain myself at the medical school where I was teaching.

So I had a brilliant idea. I could write science essays for the *Journal of Chemical Education.* They would count as papers, would involve no research, and would be fun. I did half a dozen.

Then in 1955, I had an even more brilliant idea. I could write precisely the same sort of essays, but in a more informal and pleasant style, for science-fiction magazines, and get *paid* for it. After all, I had a wife and child and a second child was on the way and I could use the money.

So I began writing science essays professionally, and it worked out well indeed. My reputation grew and I began to be asked by magazines that were *not* science fiction (and that paid better than science fiction) to write essays for them, and I agreed.

I include here, therefore, a baker's dozen (that's thirteen) recent essays that I have done on astronomy, physics, chemistry, and biology for your benefit and, I hope, amusement.

# 17
# Giant Jupiter

Earth is one of five worlds that circle fairly close to the Sun, and that make up the "inner Solar System."

Mercury, Venus, and Earth are made up of rocky materials wrapped about a large ball of nickel-iron. Mars and the Moon are made up of rock only.

Far out beyond Mars, however, lie four planets that are totally different from Earth and its neighbor worlds. They occupy the "outer Solar System," and their names are Jupiter, Saturn, Uranus, and Neptune.

These four planets are giants, each one being much larger than Earth. In fact, taken together, these planets make up 99.5 percent of the mass of all the objects that circle the Sun. The Earth and the other worlds of the inner Solar System, plus the various satellites, asteroids, meteors, and comets, all put together, make up the other 0.5 percent.

The four giant planets are totally different in structure and chemical makeup from the Earth and its companion worlds. There may be balls of rock at the center of the giants, but the outer regions are made up of normally gaseous materials under so much pressure that they are squeezed into liquids that are very hot—that are at thousands of degrees of temperature.

These giants are so different from the nearby worlds that scientists are naturally curious to know as much about them as possible. Unfortunately, they are so far away from the Earth that it is difficult to make out much in the way of details.

Of these four enormous planets, by far the largest and the most unusual is Jupiter. It is 2.5 times as massive as the other three giants put together. It is 318.4 times as massive as our Earth. Its diameter across its equator is 142,900 kilometers, as compared with Earth's 12,757.

Jupiter's visible surface is just a cloud layer floating on an enormous atmosphere of hydrogen and helium. However, if we imagine this as a real surface, it is 125 times as large as Earth's. If we imagine Earth's surface spread out on Jupiter, it would take up as much space as India and Pakistan do on Earth's surface.

If there were just one giant planet we could study in detail, it would surely be Jupiter, and it just happens that it is the one that is closest to us. Of course, even the "closest" isn't very close. Jupiter never comes closer to Earth than about 630 million kilometers. Jupiter is thus 1650 times as far from us as the Moon is; it is sixteen times as far as Venus is at its closest, and twelve times as far as Mars is at its closest.

What's more, we can't travel to Jupiter in a straight line. Both Earth and Jupiter are traveling in nearly circular orbits about the Sun. That means a rocket ship taking off from Earth is also moving in an orbit about the Sun, and it must take up the kind of orbit that, starting at Earth, curves outward until it intersects Jupiter's orbit at a point where the planet happens to be. The length of the curve is considerably greater than is the straight-line distance.

Even so, human beings have so far sent out four unmanned rocket "probes" to Jupiter—*Pioneer 10, Pioneer 11, Voyager 1,* and *Voyager 2.* Each one had to move through space for nearly two years before reaching its destination. The first arrived at the neighborhood of Jupiter in December 1973, and the fourth did so in July 1979. Each probe was equipped with cameras that could send back photographs of the planet and its satellites, and the instruments that could send back other information.

The result is that we know much more about Jupiter now than we did before the probes made their trips.

Can we now expect that spaceships with human beings aboard may someday approach Jupiter and attempt to find out even more about it than instruments under remote control can?

It would be a long trip—perhaps four years, there and back, but this is not unthinkable. The first circumnavigation of the Earth took three years, and advances in rocketry may well shorten the trip to Jupiter in decades to come.

Of course, even if we reach Jupiter, there are some things we are not likely to be able to do in the foreseeable future.

For instance, we cannot land on Jupiter's "surface," because as I said earlier, it is not a surface in the usual sense of the word at all. It is merely a cloud layer. A ship trying to land on it would simply sink deeper and deeper into the planet's atmosphere.

The temperature of Jupiter at its visible cloud layer is only -135 degrees C, which is not surprising, considering that Jupiter is five times as far from the Sun as we are and receives only 4 percent as much light and heat as we do. However, as the ship sank below the visible surface, the temperature and pressure would quickly rise and the ship would quickly be destroyed.

Of course, we might imagine that we could avoid landing on Jupiter simply by placing the ship into orbit about the planet just above the cloud layer. Jupiter, because of its great mass, has a far more intense gravitational field than Earth does; but, if the ship is in orbit, it is in "free fall"—and Jupiter's gravitational field would not be felt as "weight."

Still, after the ship had made all the observations it needed, how would it get away? Just above the cloud layer, Jupiter's gravitational pull is 2.5 times that of Earth's at its surface. In order for a ship to escape Earth's pull at its surface it has to attain a speed of 11.3 kilometers per second. To escape from Jupiter's pull at its cloud layer, a ship would require a speed of 60.5 kilometers per second. The energy requirement for attaining this speed would be enormous, and the problem of having the ship carry enough fuel to supply the energy would be more enormous still.

There are two other difficulties to going into orbit just outside the cloud layer. Above the cloud layer there are still thin wisps of gas, wisps that are thick enough to present some resistance to the motion of the ship. Such resistance would cause the ship's orbit to decay, sending it plunging into the atmosphere.

Worse yet, the probes have shown that Jupiter has a magnetic field twenty to thirty times more intense than Earth's. This field accumulates so many charged subatomic particles that Jupiter's radiation intensity can be hundreds of times that which would suffice to kill human beings.

But then, if we can't land on Jupiter or move into orbit above it at too close a distance, can we perhaps land on one of its numerous satellites and use that as a base for which to observe Jupiter?

Jupiter has sixteen satellites. Four of them are very small—less than a hundred kilometers in diameter—and are quite close to Jupiter. (Three of them were discovered by the probes.) There is also a thin ring of small particles circling close to Jupiter. All these objects are too dangerously close to Jupiter to approach, however.

Farther out are four large satellites. They are, in order of increasing distance, Io, Europa, Ganymede, and Callisto. Io is just about the size of our Moon; Europa is a bit smaller than our Moon; Ganymede and Callisto are a bit larger.

Ganymede, the largest of the four, is 5270 kilometers in diameter and is the largest satellite in the Solar System. It is, in fact, larger (but less massive) than the planet Mercury. Mercury is composed of rock and metal and is therefore more massive than the larger Ganymede, which is composed of rock and ice.

These satellites are far enough from Jupiter to be well outside its atmosphere, and they have no atmosphere of their own. Ganymede is, for instance, 1,070,000 kilometers from Jupiter, nearly three times as far as the Moon is from the Earth. Jupiter's gravity at that distance is not dangerous, and neither is Ganymede's own relatively small gravitational pull.

However, there is still Jupiter's magnetic field. It is wide enough to spread out over the large satellites and to represent a continuing danger. Callisto, the farthest of the large satellites, has the best chance of being safe in this respect, but even in this case, remaining there might be a long-term risk.

Out beyond Callisto, however, are eight small satellites of Jupiter that are probably captured asteroids. Three of them circle Jupiter at distances that average 11 to 12 million kilometers, and the remaining five are at distances that average 21 to 24 million kilometers. All of them lie far outside Jupiter's magnetic fields and offer themselves as convenient space stations.

These small satellites are far from Jupiter, but not *too* far. From those outer satellites, Jupiter looks about as large as our Moon does to us. From those satellites, which have no atmosphere, a telescope will show us at least ten thousand times the detail the same telescope would show from Earth.

Besides this, from space stations on outer satellites, probes could be sent to the inner satellites and to Jupiter itself, frequently and easily. Right now, plans are being made (though they have been delayed by the *Challenger* disaster) to send a probe into Jupiter's atmosphere. This mission can be accomplished much more efficiently from Jupiter's outer satellites.

A probe sinking into the atmosphere could not only detail temperature, pressure, and other properties as they change with depth, but between the very frigid uppermost layer and very hot deeper layers they must pass through a region of moderate temperatures. At those moderate temperatures there may be present liquid water. Indeed, it is even conceivable that a form of life might exist there, floating in the atmosphere, shifting from downdrafts to updrafts and back again in order to stay in the mild layer.

Each of the large satellites would also be probed. Some unmanned probes could be placed in orbit about each one, while others might actually be landed on the surface. Each satellite has its own points of interest. The closer they are to Jupiter, for example, the greater the tidal effects of Jupiter's gravitational pull and the greater the heat generated as a result in the satellite.

Callisto, the farthest from Jupiter of the large satellites, is about half ice and half rock. It is laden with craters from meteoric impacts during its early history. Jupiter's great distance means that Callisto received little heating effect and has been mostly unchanged in the last four billion years. Ganymede, the next farthest from Jupiter, is also icy but is less cratered. Jupiter's stronger effect on this satellite has produced changes that have led to the existence of mountain ridges and valleys.

The innermost satellite, Io, has been so strongly heated that it is completely dry. Its interior is so hot, in fact, that there are active volcanoes on its surface, the only active volcanoes known to exist in the Solar System outside the Earth. The Io volcanoes belch out sulfur, turning the entire surface of the satellite yellow and orange and filling in almost all the craters that have existed.

The most interesting satellite is Europa, the smallest, which lies between Io and Ganymede. It is covered with a smooth, icy glacier that covers the entire satellite. Meteoric impacts crack the glacier, but don't form craters, possibly because Jupiter's heating effect keeps the lower layers of water liquid, forming the only water ocean known to exist in the Solar System outside Earth. The liquid water would well up into broken areas of the glacier and freeze again.

It would be very interesting to send a probe into Europa's glacier, in an attempt to crack it and penetrate to the liquid water below. We can't help but wonder if some forms of life may have developed in this concealed ocean.

One thing we *can* be sure of. If we can reach Jupiter and its satellites and study them in great detail, we are bound to discover all sorts of interesting and unexpected phenomena.

# 18

# Pluto, The Constant Surprise

At the beginning of the century, some astronomers thought there might be a planet beyond Uranus and Neptune to account for tiny discrepancies in their motions. Perhaps some gravitational pull from a distant planet wasn't being allowed for.

Jupiter is 318 times as massive as Earth, and Saturn, which is farther away, is only five times as massive as Earth. Beyond them are Uranus, which is fifteen times as massive as Earth, and Neptune, which is seventeen times as massive. If a still more distant planet existed, it would be smaller than any of these, but it might still be six or seven times as massive as Earth.

For a whole generation astronomers tried to figure out where in the sky this planet, which produced observable gravitational effects, was. Finally, in 1930, a young American astronomer, Clyde Tombaugh, found it. He named it Pluto, because it was so far in the outer darkness that it seemed fitting to name it for the ancient god of the underworld.

But then came the first surprise. It was such a *dim* world. Astronomers thought that if it were as large as they expected, and as distant, it would be of the tenth magnitude, but it was of the fourteenth magnitude. It was only one-fortieth as bright as it should have been. Why? Perhaps it was farther away than scientists thought it was. They watched as it slowly moved across the sky; and from its speed and the direction of its motion, they calculated its orbit. It takes Pluto 247.7 years to go once about the Sun, so it's average distance is about 5,900 million kilometers from the Sun. This makes it one and one-third times farther away than Neptune, which till then was the farthest planet known; but that still leaves Pluto rather *closer* than had been expected.

In order to account for its dimness, then, scientists had to assume it was smaller than they had expected. Perhaps it was no bigger than the Earth.

Then more surprises turned up. As its motions continued to be watched, it became plain that Pluto had an orbit that was more elliptical than that of any other planet. In half of its orbit it is much farther from the Sun than it is in the other half. At its farthest point (aphelion), Pluto is 7,375 million kilometers from the Sun. At its nearest point (perihelion), Pluto is only 4,425 million kilometers from the Sun—actually a little closer to the Sun than Neptune is.

When Pluto was discovered, it was slowly approaching perihelion. In 1979 Pluto crossed Neptune's orbit and will remain inside it for twenty years. Right now it is *not* the farthest planet from the Sun; Neptune is. However, by the year 2000, Pluto will pass beyond Neptune and go sailing outward. It will then remain the farthest planet for another 220 years, when it approaches its perihelion again.

Does this mean that Pluto might some day collide with Neptune? No, because Pluto's orbit is not in the same plane as the other planets. It is inclined to the plane of the Earth's orbit by 17 degrees. If you made a small but exact model of the Solar System and the planetary orbits, which would all fit inside a pizza box, Pluto's tilted orbit would stick out of it. As a result, when Pluto moves in closer to the Sun than Neptune is, or moves out farther, is never less than 1,300 million kilometers above or below Neptune.

And what is Pluto made of? If it is a small planet, it can't have a deep atmosphere and a thick cloud layer, as the giant planets do. Instead it must be made of rock, or ice, or a mixture of the two. Either way, there might be light and dark rock, or frost-covered and bare rock. Different parts of it might be differently bright.

Sure enough, in 1954, Canadian astronomer Robert H. Hardie and co-worker Merle Walker measured the brightness very precisely and found that it went slightly up and down every 6.4 days. They decided this meant that Pluto rotated on its axis once every 6.4 days and that one hemisphere was a little darker than the other.

Meanwhile new efforts were being made to determine just exactly how big Pluto might be. Perhaps with new and bigger telescopes, Pluto might be magnified to a little orb that could be measured directly. In 1950, the Dutch-American astronomer Gerard Peter Kuiper tackled the task with a new giant telescope at Mt. Palomar, one with a 508-centimeter mirror. He did get a tiny orb, and, measuring it as best he could, he found that its diameter was about 6,100 kilometers. This meant it was only half the diameter of the Earth and, indeed, a little less than the diameter of Mars.

Astronomers were thunderstruck at this. They couldn't believe that Pluto was *that* small.

As it happens, there is another way of determining the size of Pluto. Every once in a while, Pluto, as it moves slowly across the sky, passes near a dim star. If it happens to move directly in front of the star (an "occultation") the star will wink out for a period of time. From that period of time we would get a notion of what Pluto's diameter is.

On April 28, 1965, Pluto was moving toward a dim star in the constellation Leo. If Pluto were as large as the Earth or even as large as Mars, then the planet would move close enough to the star to hide it with its rim. However, as Pluto passed in front of it, there was *no* occultation. Pluto's little sphere missed it altogehter. In order for Pluto's sphere to miss it, it had to be no more than 5,790 kilometers across; and it might well be less than that. That made it seem that Pluto was considerably smaller than Kuiper had thought. Its size would, *at most,* be halfway between that of Mercury and Mars, the two smallest known planets.

Even if Pluto were made of rock, it would have a mass of only one-sixteenth that of Earth—and perhaps less.

Then came a new development that was totally unexpected, the biggest surprise yet that Pluto had offered.

In June 1978, an American astronomer, James Christie, was studying photographs of Pluto taken under excellent conditions. Christie studied the photographs under strong magnification, and it seemed to him there was a bump on Pluto.

He looked at other photographs under magnification and they all had the bump. What's more, Christie noticed that the bump wasn't in the same place from picture to picture. By studying every picture he could get he found that the bump was moving about Pluto in 6.4 days, the period of Pluto's rotation.

Either there was a huge mountain on Pluto, or else Pluto had a nearby satellite. In 1980, a French astronomer, Antoine Labeyrie, working on top of Mauna Kea in Hawaii, showed there was a space between Pluto and the bump. Pluto might be a small world, but, to the amazement of astronomers, it had a satellite!

Christie named the satellite Charon, after the ferryman who, in the ancient myths, carried the shades of the dead across the Styx into Pluto's underground kingdom.

In 1980, Pluto passed close to another star. Pluto did not occult the star, but Charon did! From this occultation, as viewed from South Africa by an astronomer named A. R. Walker, it turned out that Charon had a minimum diameter of 1,170 kilometers (only one-third the diameter of the Moon.)

By measuring the apparent distance between Charon and Pluto, and knowing the distance of the two from ourselves, astronomers calculated that Charon was 19,700 kilometers from Pluto. (This is only one-twentieth the distance of the Moon from the Earth.)

From the rate at which Charon turned about Pluto at this distance, it was possible to calculate the mass of Pluto, and it turned out that Pluto was, at most, only about 1/500 the mass of the Earth. In fact, it was, at most, only a little over one-sixth the mass of the Moon. As for Charon, it had about one-tenth the mass of Pluto. Pluto was smaller than anyone had expected—once more it had surprised the astronomers.

Pluto was now understood to be so small that it could not be made up of rock. If it were of rock, it would be too small to have the brightness it seems to have. Rock would not reflect enough light. Pluto must be an icy body. Ice is bulkier than rock and reflects more light. An icy Pluto would therefore be large enough and reflective enough to show the brightness it does.

The latest measurements make Pluto smaller still. We now consider Pluto to be only 2,280 kilometers across, and it has only one-tenth the mass of our Moon (or 1/800 the mass of the Earth). Charon is only 1,290 kilometers across and has only 1/100 the mass of our Moon.

The Pluto-Charon combination breaks a record. Usually, a planet's satellites are much, much smaller than the planet itself. Ganymede, Jupiter's largest satellite, has only 1/10,000 the mass of Jupiter, for instance. Our Moon, however, is 1/81 as massive as the Earth. The Moon is a very large satellite for a planet as small as Earth to have, so that Earth-Moon was the closest thing to a "double planet" our Solar System seemed to have. But Charon is one-tenth as massive as Pluto. Pluto-Charon is much closer to a double planet than Earth-Moon is.

Charon circles Pluto in such a way (as seen from Earth) that every 124 years, for a five-year period, it moves in front of Pluto, then behind it. It goes through this period of eclipse when Pluto is farthest from the Sun and again when it is nearest.

Charon happened to be discovered just before it began its five-year period of eclipses, so that astronomers are now watching the effects avidly. What's more, Pluto is now at perihelion and is now closest to the Sun and to us—just when it can be studied best. If Charon had been discovered only fifteen years later, astronomers would have lost their chance and would have had to wait two and a half centuries for the next near-point eclipses. (By then, though, we would surely have sent rockets past Pluto.)

When two worlds are close together, tidal effects slow their rotations. Thus, the Earth's tidal effect has slowed the Moon's rotation to the point where it shows only one hemisphere to the earth as it circles us. Earth's rotation is also slowing because of the Moon's tidal effect, but Earth is so large that the slowing effect has only been partial so far.

Pluto and Charon, however, are so close together that the tidal effect on each is magnified. The two worlds are, in effect, so small that they are easily and quickly slowed. The result is that both worlds have been slowed to the point of showing only one hemisphere to the other. They face each other permanently and turn about each other as though they were all one piece. They are the only two worlds in the Solar System that turn about each other in this fashion.

It is possible to study the infrared light that is reflected by Pluto and Charon. When Charon is behind Pluto, we see only the reflected infrared of Pluto. When Charon comes out from behind Pluto, we see the reflected light of both, and if we subtract the reflection of Pluto, we get the reflected light of Charon only.

From this reflected light, astronomers in 1987 worked out the chemical nature of that world.

Thus, they have discovered that the surface of Pluto seems to be rich in methane, a substance that, on Earth, is a major part of the natural gas we use as fuel. Methane freezes at a very low temperature so that even at Pluto's temperature, which may be –240 C, some of it will still evaporate and become a gas. It would seem, then, that Pluto has an atmosphere of methane gas that is about 1/900 as dense as Earth's atmosphere (and nearly a tenth as dense as the thinner atmosphere of Mars).

Naturally, the temperature is lower at Pluto's poles, so there is more frozen methane there. Pluto may, therefore, have polar icecaps of frozen methane, which become larger as it moves farther from the Sun.

Astronomers were surprised to find that Charon's reflected light was quite different from Pluto's. Since Charon is smaller than Pluto, it has a smaller gravitational pull. It can't hold on to the molecules of gaseous methane as well, and the methane has escaped from Charon during the billions of years that the Solar System has existed.

What's left behind on Charon is frozen water, which doesn't vaporize at the temperatures out there and, therefore, isn't lost. Just the same, Charon may have an atmosphere, too. Pluto's thin atmosphere spreads out widely under Pluto's small gravity, which can't keep the atmospheric gas close to the planetary surface. The methane therefore extends beyond Charon's orbit, so that the two worlds may have a common atmosphere. This means Charon is moving through the methane, and the gas resistance is very gradually slowing it down and bringing it closer to Pluto. Eventually, the two will crash together and coalesce, and that would, perhaps, be the final surprise.

Except that we can't trust Pluto. It may have more surprises up its sleeve. For instance, it is far too tiny to produce some of the gravitational effects we've observed on Uranus and Neptune. What does, then? Is there a larger and still farther planet out there somewhere?

# 19
# A Hole in the Sky

Back in the early 1800s an English astronomer, William Herschel, noticed a dark patch in the Milky Way. All about it there blazed innumerable stars, but within the patch there was nothing. Herschel was amazed. He thought this was a region in which no stars existed, a kind of tunnel through the stars the opening of which just happened to be pointing at us.

"Surely," he said, "there is a hole in the heavens."

But as the years passed, other such dark patches were found and it became unbelievable that so many tunnels should happen to have their mouths pointed toward us. Instead, astronomers decided (correctly) that these patches were clouds of gas and dust that blocked the light of stars on the other side. (This is similar to the way black smoke from a large fire might obscure objects beyond it.)

The thought of holes in the sky vanished for a while. Eventually, astronomers realized that the stars in the sky formed a lens-shaped body called the Milky Way Galaxy. It contains at least 200,000,000,000 stars, which are scattered all through it—along with occasional clouds. There are no particular holes.

In the 1920s, however, it became clear that the Milky Way Galaxy was not the only one there was. There were other galaxies, most of which were smaller than ours, but some of which were larger. The nearest large galaxy, the Andromeda Galaxy, is 2,300,000 light-years away from us and probably contains twice as many stars as our galaxy does. (Its distance is such that light, traveling at a steady 186,282 miles per second, would take 2,300,000 years to go from our galaxy to the Andromeda. That is what is meant by 2,300,000 "light years.") In fact, our galaxy is a member of the "Local Group," containing about two dozen galaxies, including the Andromeda. The Local Group is about 3,000,000 light years across.

Beyond the Local Group are still other galaxies, millions of them—billions of them. Other groups, or "clusters," of galaxies exist, some of them much larger than the Local Group. Some clusters of galaxies contain thousands of members.

If one looks casually at galaxies through a telescope, they seem to occur everywhere in the sky (where they aren't obscured by clouds of dust or by particularly thick crowds of nearby stars). At first, it was natural to think that they were scattered evenly through space, just as stars are scattered more or less evenly through the Galaxy.

However, once astronomers learned how to determine the distances of individual galaxies, they could make three-dimensional models in which different galaxies were placed in the correct direction and at the correct relative distances. It then turned out that galaxies existed in curved lines and planes, so that the Universe looked like a mass of soap-bubbles. The galaxies are concentrated along the bubble films or boundaries, and inside the bubbles there is little or nothing—that is, "voids." These voids are like holes in the sky. They are holes like those that Herschel *thought* he had seen, but they are apparently real holes and much, much larger than Herschel could possibly have imagined.

The largest void yet discovered (first noted in 1981) is located in the constellation of Bootes (boh-OH-teez) the Herdsman. It is therefore called the "Bootes void." It is about 600,000,000 light-years away from us, which makes it 260 times as far away as the Andromeda Galaxy.

The Bootes void is roughly spherical and is 300,000,000 light years in diameter. It is hard to imagine how enormous the void is. This volume of almost-nothingness is over 5,000 trillion times the volume of our own galaxy. We can therefore imagine many trillions of galaxies poured into such a void if they were tightly squeezed together. Of course, galaxies *aren't* tightly squeezed together. They exist far apart, hundreds of thousands of light-years apart on the average. Even so, the Bootes void could easily contain over a thousand galaxies that are scattered in the normal manner.

If the Bootes void contained nothing at all, that would be interesting enough, but there are a few objects that, in 1987, were observed in the void. They are "emission-line galaxies," galaxies that are particularly energetic and heated to high average temperatures. Ordinarily, such galaxies are not common. Only about one out of every fifteen to twenty galaxies are "emission-line." And yet all of the few galaxies that have been spotted in the Bootes void belong to this rare class.

This creates a fascinating problem for astronomers and makes for a puzzle within a puzzle within a puzzle. Here's how it works—

When the Universe first formed in the Big Bang, about fifteen billion years ago or so, it was a tiny object that is thought to have been homogeneous; that is, it had no irregularities in it. (Scientists have not thought of any way of accounting for any irregularities at the start.)

It would seem, then, that as this tiny object expanded equally and very rapidly in all directions, it would continue to stay homogeneous. And yet it didn't. Instead of forming a Universe containing a uniform and evenly spread out quantity of matter and energy, it broke up into huge lumps that developed into galaxies and clusters of galaxies.

Astronomers have not yet reached a clear decision as to how galaxies could form in the first place. Some think that perhaps black holes formed here and there in the early days of the Universe and these served as nuclei about which the galaxies gathered—but that's still just a guess.

Then, even if you simply assume that galaxies formed, why did they form in soap-bubble manner, leaving voids of various sizes? That's the puzzle within the puzzle. Astronomers don't know. Possibly there were "small bangs" after the Big Bang, forming bubbles and voids, but that's just a guess.

Finally, a new question has now arisen. Even if you simply assume that galaxies formed in bubbles, and left voids, why are there only emission-like galaxies in the biggest void? What does that mean? That is the puzzle within the puzzle within the puzzle. And so far, of course, astonomers don't have a clue.

# 20

# Our Changing Perception of the Universe

In 1900 the stars we saw in the sky by eye and by telescope were thought to make up a huge pancake-shaped cluster, which was called the Galaxy. The most daring estimate of its size made it about 20,000 light-years across (one light-year is equal to 5,880,000,000,000 miles). The Galaxy was thought to contain two or three billion stars.

Astronomers were of the opinion that the Galaxy represented the *entire* Universe.

To be sure, that was large enough, considering the long centuries during which it was thought that the Solar System was almost all there was, and that it was surrounded by a thin shell of a few thousand stars. Nevertheless, advances since 1900 quickly dwarfed this apparently "large" picture.

For one thing, astronomers learned new ways of measuring the distances of stars: By 1920 the true dimensions of the Galaxy were worked out by men such as Harlow Shapley. It turned out that the Galaxy is 100,000 light years across and the number of stars it contained amounts to 200 or even 300 billion. The Galaxy was a hundred times as large as it had seemed to be only twenty years before.

What's more, this monstrously enlarged Galaxy was *not* all there was to the Universe.

There were certain small cloudy patches in the sky, the so-called "nebulae," that glowed but didn't seem to contain stars. Were they possibly so far away that the stars they contained were too small to be seen individually even by very good telescopes? If so, they would have to be enormous conglomerations to seem as bright as they do. They would be other galaxies. During the 1920s the astronomer Heber D. Curtis produced evidence in favor of considering the nebulae to be other galaxies, and eventually astronomers were convinced.

For the first time, astronomers recognized the Universe to be what it now seems really to be: a collection of galaxies, each one of which is made up of anywhere from a few billion to a few thousand billion stars.

What's more, the astronomer Edwin P. Hubble was able to demonstrate quite convincingly, toward the end of the 1920s, that the Universe was not static. The galaxies existed in clusters that were all separating from each other so that the distance between them was growing steadily greater. In other words, the Universe was expanding.

Hubble and other astronomers worked out ways of determining the distances of the other galaxies. Even the closer ones were millions of light-years away. By the 1950s, some very dim galaxies were detected that were nearly a billion light-years away.

Then, in the 1960s it was found that certain objects that had seemed to be dim stars in our own galaxy were really very far away. They were called "quasars"—galaxies that are so distant that only their very bright central region can be seen, shining like a star. Even the nearest quasar is at least a billion light-years away, and by now some quasars have been detected that are at least ten billion light-years away.

If we compare the astronomical situation of 1989 with that of 1900, it would seem that the Universe, as viewed now, is a million times as wide as it was thought to be then. Instead of the single Galaxy astronomers knew in 1900, it is thought now that there may exist as many as a hundred billion of them.

* * *

How old is the Universe?

In 1900, astomomers had no idea. Perhaps the universe had existed forever, or perhaps it had been created a few thousand years ago by some divine action. There seemed absolutely no way of telling by studying the stars.

Once it was found that the Universe was expanding, however, it was clear that if we looked back in time, the Universe would seem to be contracting. If we looked far enough back into the past, the Universe would be seen to have contracted to a very small size, and that would represent its beginning.

The astronomer Georges E. Lemaitre first suggested this in the late

1920s. He felt that a very small object had once exploded to form the Universe and that the Universe was still expanding today because of the force of that explosion. In the 1940s the physicist George Gamow called this explosion the "big bang," and the name stuck.

When did the big bang take place?

That depends on how far away the galaxies are and how quickly they are separating. Once those figures are known, astronomers can calculate backward and see how long ago all the galaxies were at a single point.

The best estimate right now seems to be that the big bang took place, and the universe was born, 12 to 15 billion years ago.

It takes light one year to travel a light-year. When we look at a very distant quasar, one that is 10 billion light-years away, the light took 10 billion light-years to get to us, and we see it as it was 10 billion years ago, near the beginning of the Universe's existence. We can't expect to see anything much farther away, for we will then approach the time soon after the big bang, when the galaxies had not yet been formed.

* * *

What are stars like?

In 1900 they were thought to be like our Sun, some larger and brighter, some smaller and dimmer, but we knew nothing else. In the 1930s, however, Hans A. Bethe worked out the nuclear source of a star's energy.

With this known, the nature of a star's evolution could be understood: how it formed; how it remained for long periods of time in stable form; how it finally began to run low on nuclear fuel, swelled to a "red giant," and finally collapsed.

In the 1910s, "white dwarfs" had been discovered, small hot stars, no larger than Earth, but with all the mass of a Sun squeezed into its small body. These came to be understood as the natural collapsed form of relatively small stars.

Giant stars explode as "supernovas" before collapsing, and then shrink to objects even smaller than white dwarfs. In the 1960s such "neutron stars" were discovered, objects no more than eight miles across that yet contain all the mass of a star like our Sun. Scientists believe that very large stars will collapse even further to still tinier objects which such strong surface gravity that nothing can escape them, not even light. These are "black holes."

Black holes are very hard to observe, but by the 1980s astronomers were confident that large black holes might exist at the center of galaxies, even our own. The presence of such black holes might account for explosive events that take place at the centers of many galaxies, making the Universe a much more violent place than any astronomer had dreamed it was back in 1900.

* * *

Astronomers are now aided by instruments that did not exist in 1900. In 1900, there were telescopes, spectroscopes, and cameras, but these all worked with ordinary light exclusively. There didn't seem to be anything else to work with.

In the 1930s, however, it was found that floods of radio waves were bombarding the Earth from the stars. In the 1950s radio telescopes were built to study and analyze these waves, and by means of such telescopes very distant objects could be studied in the kind of detail that would be impossible for ordinary telescopes. Quasars, neutron stars, black holes, and other objects might never have been discovered without radio telescopes.

By 1900 no man had ever left the ground in powered flight. Only balloons existed. In that year, however, the first dirigible flew, and in 1903, the first airplane. In the 1920s the first liquid-fueled rocket was launched. In the 1950s the first artificial satellite was put into orbit. And in 1969 human beings first stepped onto the soil of a world other than the Earth—the Moon.

Meanwhile rockets and probes began to enlarge our vision of our Solar System past the imaginings of astronomers of 1900.

The Moon was photographed at close quarters and mapped in detail—not only the side we see, but the far side, which until the late 1950s had never been seen.

Mercury, Mars, and the two Martian moons were also mapped. Mars was shown to have no canals (which, in 1900, were thought to exist by some astronomers), but it did have craters and dead volcanoes. Even Venus was mapped, right through its clouds, by radar.

Probes, travelling very far from Earth, took close-up photographs of Jupiter and Saturn, showing unexpected details of Saturn's rings. Distant satellites were explored. Io has active volcanoes, Europa a covering of smooth ice, Titan a very thick atmosphere. Other, small satellites were discovered.

* * *

And where will we be by the end of the twentieth century?

Astronomers hope that space exploration will continue and that in the course of the next decade many more surprising things about our Solar System will be discovered.

In addition, astronomers hope soon to place into space a large telescope that will be able to view the Universe without the interference of Earth's atmosphere. This telescope should be able to observe distant objects in far greater detail than is now possible and to tell us, perhaps, whether the universe will expand forever . . . or whether it will begin to collapse again one day. It may also make it possible to learn in greater detail just how the Universe evolved.

There have been great times for astronomy in this century, and there are even greater times ahead.

# 21

# What Is the Universe?

The word *universe* comes from Latin words meaning "turning as one." It is everything, treated as a unit. It is all the matter and energy that exists.

We have the disadvantage of studying the Universe from within. We can see those portions near to us, but the farther portions become progressively dimmer, progressively blurred. Even with all our instruments, much of the Universe remains too distant and dim to see at all—let alone in detail.

From what we see, though, we can come to conclusions. So suppose we imagine we are viewing the Universe from outside under such conditions that we are aware of the whole thing. (This is impossible, of course, since there is no such thing as "outside the Universe," but let us imagine it anyway.)

The Universe would look like a three-dimensional mesh of fine strands of light, with empty spaces between. There would be a great many small empty spaces, a smaller number of larger ones, a still smaller number of still larger ones. As for the lines of light, they would gather here and there in small knots or clumps of light, with a smaller number of brighter knots, and so on.

The Universe would most resemble a sponge built of light. The curving lines and sheets of light are built up of about one hundred billion dots of light (some considerably brighter than others). Each of these dots is a galaxy.

The Universe as we view it would be most notable for its stillness. Nothing whatever would seem to be happening to it. The reason for this is that no progressive change, large enough to be noticeable under our Universal view, can possibly take place at faster than the speed of light. The speed of light (186,282 miles per second) may seem unimaginably fast to us, but on the scale of the Universe as a whole, light may be considered virtually motionless.

Suppose, for instance, that as a result of some unimaginable event, the central point of one of the galaxies of the Universe ceases to emit light. It grows dark. Suppose that a wave of such darkening spreads outward from that central point in all directions at the fastest possible speed, that of light. We, watching from without, might see the galaxy (visible to us as a dot of light) begin to grow slightly dimmer, but it would take tens of thousands of years before the galaxy would blank out completely. It would take hundreds of thousands of years for the darkening to extend to other, neighboring dots. It would take some 12 *billion* years, at the very least, for the entire Universe to darken.

If we began watching at any stage of this Universal darkening we would see absolutely no change in the course of a lifetime, and very little in the course of a hundred lifetimes. (The same would be true, by the way, if the Universe were dark to begin with and began to grow light from some central point, the influence spreading outward at the speed of light.)

We, ourselves, are as much a prisoner of our place and time as everything else is. We cannot, under any circumstances we know of, go faster than the speed of light. At that speed, it would take us about 160,000 years to go to the far end of our own galaxy and back, and 4,600,000 years to travel to the Andromeda galaxy, our nearest large neighbor, and back. To be sure, at the speed of light, Einstein's relativity tells us, the rate of time-passage will sink to zero and, for us as we travel, no time will seem to be passing. Back on Earth, however, when we return to it, we would find that 160,000 years had irrevocably passed while we were visiting the far end of the Galaxy or that 4,600,000 years had passed while we were flashing by the Andromeda and returning.

It is not likely that we will be able to go at the speed of light, however. The greatest practical speed may prove to be no more than a fifth the speed of light, in which case the relativistic slowdown of time for the traveller is insignificant. It would then take 800,000 years of the astronaut's real time to visit the other end of the galaxy and return and 23,000,000 years to visit the Andromeda and return.

It may be, then, that with the best will in the world, any man in his own lifetime may be able to do no more than visit the very nearest stars, and from the Universal view that travel distance will be essentially zero.

Suppose, though, that, as we view the Universe as a whole, we overcome its motionlessness, by imagining that we speed up time a millionfold. Or, alternatively, we can imagine that some kindly superbeing has taken a detailed photograph of the Universe every hundred thousand years and that now we have the opportunity to run the film through a projector at the usual sixteen frames a second.

At this speed, the galaxies undergo rapid changes. Each one spins rapidly about its center. If it is a spiral, the spiral arms may disappear and reappear. None of these changes would be visible from our Universal view, of course. The dots of light would remain just dots of light.

At this speed, also, some galaxies will be exploding in a sudden burst of light, some will develop black holes that will grow enormously and devour millions of stars in a matter of seconds. Other galaxies will collide and produce incredible showers of radio waves and other radiations. None of this will be particularly visible either. Some of the dots of light in our Universal view may brighten slightly and others may dim slightly, but we probably won't notice it happening without careful measurements, if then.

In that case, will even speeding up time do nothing to remove the Universe's changelessness? Not so. There is one change that is the overwhelming fact about the universe.

As we watch the film run we will notice that the Universe is visibly *expanding*. The holes in the spongy structure will slowly grow larger, and the curves and swoops of light will slowly thin out and spread apart, so that the intensity of light in any one spot will dim. In short, the Universal sponge will grow larger and larger and dimmer and dimmer.

We might also run the film backward. In that case the Universe will be visibly *contracting*. The holes in the spongy structure will slowly grow smaller, and the curves and swoops of light will slowly thicken and tighten. In short, the Universal sponge will grow smaller and smaller and brighter and brighter.

If we continue to run the film in the normal direction indefinitely, the Universe may expand and dim indefinitely until it is too dim to see at all. If, however, we continue to run the film in the backward direction, there is a limit to the length of time we can continue to do so, for eventually the Universe must shrink to nothing.

In fact, if we start at the presesnt and run the film backward at 100,000 years every one-sixteenth of a second, then in about two hours the Universal view will be seen to have contracted itself into a tiny dot that is unbearably bright (though not in visible light) and unbearably hot, and then it will blink out into nothingness.

If we start at that point of nothingness and run the film forward, the dot will appear with its unbearably bright heat and quickly expand and cool. That is the "big bang," in which, astronomers now suspect, all the matter and energy of the Universe was formed out of nothing, in accordance with the peculiar rules of quantum theory.

This big bang presents astronomers with a fascinating problem. At the moment of the big bang, that original point of light must have been homogeneous. Everything in it must have been completely mixed. As it expanded, it should have stayed completely mixed. The whole Universe today ought to be just one large, ever-expanding, ever-thinning gas, which would always be the same everywhere in the Universe.

Instead, from the Universal view, we see a terribly uneven Universe. Matter and energy have coagulated into the dots we call galaxies, and these have, in turn, collected into lines and curves of light that give the

Universe a spongy appearance. How can the Universe have gone from a featureless dot of light to a sponge? Cosomologists are still arguing over it and trying out various theories.

Another problem is this: Will the Universe expand forever?

The Universe is expanding against the pull of its own gravity, and, as a result, its rate of expansion is slowing. But is this braking effect of gravity sufficient to bring the expansion to a complete halt some day and start a contraction instead?

That depends on the quantity of matter in the Universe, for matter is the source of the gravitational pull. At the moment, it seems that the amount of matter we can detect is not more than about 1 percent of the quantity needed to stop the expansion someday. Yet there are some indications that the expansion *will* stop some day. If that is so, it means that there is at least a hundred times as much as in the Universe than we can detect so far.

This is called "the mystery of the missing mass," and cosmologists are arguing over this heatedly.

# 22

# The One-Man Revolution

If Albert Einstein had been alive on March 14, 1979, he would have been celebrating his hundredth birthday. He would also have been looking at a world of science that had been revolutionized as a result of his work.

He was born in 1879 in Germany, and there was no sign in his younger years that he was to be a one-man intellectual revolution. He showed no particular promise as a youngster. As a matter of fact, he was so slow in learning to speak that there was some feeling that he might prove retarded. And he did so badly in Latin and Greek in high school that he was invited to leave by a teacher who said to him, "You will never amount to anything, Einstein."

He managed to get into a Swiss college—with difficulty. He managed to graduate—with difficulty. He could find no teaching post, and in 1901, thanks only to the influence of the father of a friend, he managed to find a position as a junior official at the Patent Office at Berne, Switzerland.

There he began his work, and for it, fortunately, he required only a pencil, paper, and his deep understanding of mathematics.

In 1905, when he was twenty-six years old, he burst upon the consciousness of the scientific world with important papers on three different subjects.

One paper dealt with the photoelectric effect, whereby light falling upon certain metals stimulates the emission of electrons. In 1902 it had been found that the energy of the emitted electrons did not depend upon the intensity of the light. A bright light of a particular kind might bring about the emission of a greater number of electrons than a dim light of the same kind, but not of more energetic ones. This puzzled physicists of the time.

Einstein applied to the problem the quantum theory worked out five years earlier by Max Planck. In order to explain the manner in which radiation was given off by bodies at different temperatures, Planck had postulated that energy came off in discrete chunks, which he called "quanta." The higher the frequency of light (and the shorter its wavelength), the more energy there was in the quanta.

The quantum theory was not generally persuasive at the time, since Planck seemed to be merely fiddling with numbers to make an equation work. Even Planck himself doubted that quanta had any real existence—until Einstein seized upon the concept.

Einstein showed that it took a quantum with a certain amount of energy to eject an electron from a given metal. Hence, light with frequency above some value would eject electrons, and light with frequency below that value would not. Very weak light of high-enough frequency would eject a few electrons; very strong light of insufficient frequency would eject none. The higher the frequency of light and the larger the quanta, the more energetic the ejected electrons.

Once the quantum theory was found to work in a totally unexpected direction, scientists had to accept it. Quantum theory revolutionized every aspect of physics and chemistry. Its acceptance marks the boundary line between "classical physics" and "modern physics," and Einstein had at least as much to do with establishing that boundary line as Planck did.

For this feat, Einstein was eventually awarded the 1921 Nobel Prize in physics. And yet the photoelectric effect was not the direction in which Einstein achieved his greatest effects.

In a second paper in 1905, Einstein worked out a mathematical analysis of Brownian motion, first observed three-quarters of a century before. It had then been found that very small objects suspended in water, such as grains of pollen or bits of dye, jiggled about randomly for no known reason.

Einstein suggested that water molecules were in random motion and that from moment to moment a few more molecules struck the small object from some one direction than from another. The suspended object was therefore driven first in one direction, then in another. Einstein worked out an equation governing such motion in which, among other things, the size of the water molecules figured.

Atoms and molecules had been part of chemical thinking for just one century at the time, but there was no direct evidence that such things existed. For all any chemist could tell, they were merely convenient fictions that made it easier to understand chemical reactions—and nothing more. Some scientists, such as F. W. Ostwald, insisted on considering atoms as fictions and strove to interpret chemistry without them.

Once Einstein's equation was published, however, it offered a chance to make a direct measurement of atomic properties. If all the values in the equation, except for the size of the water molecule were determined, then the size of the water molecule could be calculated.

In 1913, J. B. Perrin did just that. He calculated the size of the water molecule. From that calculation the size of other atoms came to be known. Ostwald abandoned his objections and, for the first time, atoms were universally recognized as real objects whose existence need not be accepted on faith alone.

Having established both quanta and atoms, Einstein might well have considered he had done enough for the year, but his greatest accomplishments were to come.

Still in 1905, Einstein published a work that established a new outlook on the Universe, one that replaced the old view of Isaac Newton, which had reigned supreme for two and a quarter centuries.

By the old Newtonian view, velocities were strictly additive. If you were on a train moving at twenty miles an hour relative to the ground, and, standing on its roof, you threw a ball forward in the direction the train was moving, with the ball traveling at twenty miles an hour relative to the train—then that ball travelled twenty plus twenty, or forty miles an hour relative to the ground. That was viewed as being as certain and as exact as the fact that twenty apples plus twenty apples were forty apples.

Einstein began with the assumption that the measured velocity of light is always constant, regardless of any motion of its source relative to the individual measuring the light.

Thus, light from a flashlight on a stationary train would move forward at a speed of 186,282 miles per second relative to the ground. If the flashlight were on a train moving forward at a rate of twenty miles an hour, the light from the flashlight would still travel forward at 186,282 miles per second relative to the ground. If the flashlight were on a train moving forward at 100,000 miles a second, or 186,000 miles a second, the light from the flashlight would travel forward—at 186,282 miles per second relative to the ground.

This seems against common sense, but what we call "common sense" is based on our experience with velocities much smaller than that of light, where the velocities are indeed additive—almost. Einstein, beginning with his assumption, worked out a formula for adding velocities that showed that even at ordinary speeds the addition was not quite arithmetical and that twenty plus twenty was not *quite* forty. The greater the speeds, the more they fell short of a simple arithmetical addition until, at the speed of light, there was no addition at all.

All sorts of peculiar-seeming consequences followed from his assumption. It turned out that nothing with mass could go faster than the speed of light in a vacuum. It turned out that length in the direction of motion decreased with velocity, that mass increased, that the rate of time passage slowed. It also turned out that light need not be viewed as a vibration of a mysterious substance called "ether." Instead, light could travel through a vacuum in the form of discrete particle-like quanta, which came to be called "photons."

All of Einstein's equations simplified to Newton's equations if the speed of light were taken as infinite. It is because the speed of light is so high that Newton's equations work well enough at low speeds such as those involved in rocketry. It is because the speed of light is not infinite that Newton's equations do not work at high speeds, such as those involving subatomic particles.

That is why Einstein's theories of relative motion had to be accepted despite their peculiar consequences. Einstein's equations worked where Newton's did not. Particle accelerators, for instance, couldn't be made to work without an understanding of Einstein's theory, and we would then have been unable to learn anything but the simplest aspects of subatomic physics without them.

Einstein's equations also showed that mass was a form of very concentrated energy and gave the equivalence of the two (the famous $E = mc^2$). It forced a reinterpretation of the conservation of energy and led to an understanding of the significance of nuclear energy. That, in turn, led in a direct line to the development of the nuclear bomb and nuclear reactors—for good or evil.

Einstein's 1905 theory applied only to the special case of uniform nonaccelerated motion relative to an observer, so it is called "the special theory of relativity." In 1915, he extended it to accelerated motion and produced "the general theory of relativity."

It was this general theory that was the foundation of modern cosmology and cosmogony. It made it possible, for the first time, to consider rationally the properties of the Universe as a whole and the manner in which it might have come into being.

The equations worked out by Einstein in connection with general relativity explained the motion of Mercury's perihelion, something that Newtonian theory had failed to explain. They predicted that light would curve in its path on passing through a gravitational field, and this was confirmed in 1919 when the position of stars near the Sun was measured during a total eclipse. They predicted that light would lose energy in moving against gravitational pull, and this was confirmed in 1925 by studying the light of Sirius's white-dwarf companion. That, incidentally, was the final proof that such things as white dwarfs could exist.

The equations of general relativity predicted that the universe would be expanding, something that was proved in the 1920s. It also predicted the existence of gravitational waves and of black holes.

In 1917 Einstein considered atoms and molecules as they gained and lost energy a quantum at a time. Einstein showed that if a molecule gained energy and was struck by a photon containing energy in an amount exactly equal to the energy gained by the molecule, that molecule would give up the energy it had gained. The molecule would emit a photon exactly equal in size to the photon that had struck, and one that was moving in the same direction besides. Where one photon had come in, two identical photons would leave.

Thirty-four years later, this principle was used by C. H. Townes to devise the maser. And, nine years later still, T. H. Maiman used it to devise the laser.

The laser, a device which produces tight beams of monochromatic coherent light, may be particularly crucial to new advances in communication in the near future, as it replaces radio waves for the purpose. Lasers may also be used to spark off hydrogen fusion and to make possible, in this way, the development of controlled fusion energy that may very well solve the world's energy crisis.

And that, too, arises from an Einsteinian concept.

There was a late contribution, too. In 1940, a group of young scientists was anxious to get President F. D. Roosevelt to supply government money for the task of developing a nuclear bomb before the Germans or Japanese could succeed in doing so. It was Einstein who was asked to sign the letter to the President. Only Einstein had the prestige to carry the day.

By that time, though, Einstein was outside the mainstream of physics. In 1929 W. K. Heisenberg had worked out the "uncertainty principle," in which it was shown that certain fundamental physical properties, such as momentum and position, could not both be measured with more than a particular definite accuracy, even in principle. Some things could only be spoken of in terms of probabilities and statistics.

Einstein could not bring himself to accept this. The notion was too uncomfortable for him. "I cannot believe," he said, "that God plays dice with the Universe."

Unfortunately, it is the principle of uncertainty that explains many important aspects of the Universe as it is, and when Einstein turned away from that, he was left behind by the advance of science. In the last third of his life he therefore accomplished little.

But it didn't matter. He had accomplished enough for a dozen men in the middle third of his life.

# 23
# The Fifth Force

There are four known forces in the Universe, four forces that cause objects to move toward each other or, in some cases, away from each other. There are four forces of attraction and/or repulsion.

The first is the "gravitational force," which holds you to the ground and will make you fall if you are not careful. The second is the "electromagnetic force," which holds atoms and molecules together and which, within the atom, holds the electrons to the central nuclei. The third is the "strong force," which holds the particles inside the central atomic nucleus together. The fourth is the "weak force," which allows some atomic nuclei to break down, producing radioactivity, and which causes the Sun to shine.

All four forces are absolutely esssential to the Universe as we know it. Without all four forces working as they do, matter could not exist, stars and planets could not exist, *we* could not exist.

Is there, however, a fifth force? Until very recently, scientists were quite convinced that there wasn't. The four known forces seemed to explain everything, and a fifth force was unnecessary.

But let's take a closer look at the four forces. They are unequal in strength. The strongest is the strong force, which is why it is called that. When two protons are placed in contact the strong force pulls them together, while the electromagnetic force pushes them apart. However, the strong force is over a hundred times as strong as the electromagnetic force, so the protons remain together and atomic nuclei can exist. The weak force is called that because it is far weaker than either the strong force or the electromagnetic force. The strong force is a hundred trillion times as strong as the weak force.

That leaves the gravitational force, by means of which the Earth not only holds you to the surface, so that you can't possibly fall off; the Earth also holds the Moon in orbit, and the Sun holds the Earth in orbit. That would lead you to suppose that the gravitational force is super-strong. It isn't! The gravitational force is far and away the weakest of the four. The strong force is about ten million trillion trillion trillion times as strong as the gravitational force.

Well, then, why do the effects of gravitation bulk so large in the Universe? The answer is that the strong force and the weak force have a very short range. Their strength falls off so rapidly with distance that they simply can't be felt at distances of as much as a trillionth of an inch. They can only be felt inside the nucleus.

The electromagnetic force and the gravitational force, however, have enormous ranges. Their strength falls off so slowly with distance that they can be felt over many light-years. The electromagnetic force, however, has both an attractive effect and a repulsive effect and the two are almost exactly balanced. On the whole, therefore, the electromagnetic force is felt only when one or the other, the attraction or the repulsion, has a very slight edge. So at great distances, it can be ignored.

The gravitational force, however, produces only an attraction. What's more, even though it is so weak, it increases with the quantity of matter (the "mass") in a body. Two rocks hardly attract each other, because they have so little mass. Even asteroids don't build up much gravitation. Something as huge as the Earth and the Moon hold together powerfully, however. The terribly weak gravitational force builds up to large quantities given that much mass. The gravitational pull of the Sun is far larger, and the gravitational pull of an entire galaxy of stars is larger still. It is therefore the gravitational force that holds the Universe together.

The mass that produces gravitation is called "gravitational mass." Mass also resists change in motion. It is easy to swat a light ping-pong ball to one side, but a ball of platinum the same size as a ping-pong ball and moving at the same speed would be far more massive and far more difficult to swat to one side. This reluctance to change motion is called "inertia," and since it increases with mass, people speak of "inertial" mass. Both the gravitational force and the inertial effect can be used to determine the mass of an object, and they always seem to give the same answer.

When Isaac Newton worked out the law of gravitation, he assumed that inertial mass and gravitational mass were always equal. So did Albert Einstein when he improved Newton's theories. Because they are equal, something massive is more "reluctant" to fall, but is pulled harder by gravitation. The two effects balance, and objects of different mass all fall at the same increasing speed.

Scientists have carefully measured the way objects fall and the way they respond to both inertia and gravitation, and it seems the two are indeed the same to about 1 part in a trillion.

Nevertheless, some scientists aren't certain. The two phenomena, inertia and gravitation, seem so different from each other that one can't help but wonder why these two different ways of measuring mass always come up with the same answer. Is it possible they don't really?

In the course of the last year or so, scientists have been making very delicate measurements, and some seem to think that gravitational mass and inertial mass are *not* exactly the same. There is a tiny, tiny discrepancy.

One way of explaining the discrepancy is to suppose there is a fifth force that is even *weaker* than gravity, a hundred times as weak. What's more, it would have a fairly short range, so that it could only be felt at distance of not more than half a mile, perhaps. Moreover, instead of being a force of attraction, pulling objects together, this new force would be one of repulsion, pushing objects apart. Finally, it would depend for its strength not only on the total mass but on the mass of particular atomic nuclei, so that its effect would be different on iron, say, than on aluminum.

All these properties are so strange that most scientists are very reluctant to accept the notion. Furthermore, the experiments involved are so delicate, and they produce effects so small, they don't seem very trustworthy. However, a number of scientists are busy devising experiments that will be still more refined; and, within another year, their results may definitely reveal whether or not there is a fifth force. If there is, scientists will have a lot of explaining to do, and things could get very exciting.

# 24

# Two At a Time

The simplest atomic nucleus is made up of only a proton. All other atomic nuclei contain neutrons as well as protons. Some 265 proton-neutron combinations are stable, and atomic nuclei made up of any of these combinations will remain as they are indefinitely, if left to themselves. All the ordinary matter of the Universe, including you and I, are made up of these various stable combinations.

There are thousands of proton-neutron combinations, however, that are *not* stable, that break down into one or another of the stable combinations. Some of these breakdowns are very fast, some are very slow, and some are at any speed in between.

These unstable nuclei are created, in the first place, by violent events, like supernova explosions, and the only ones produced in this way that exist on Earth are those that break down, once formed, only very slowly indeed. It takes billions of years for sizable amounts of uranium or thorium to break down, for instance, which is why there is still uranium and thorium atoms in the Earth's crust now.

Some unstable nuclei are being manufactured continually by cosmic rays or by bombardment of ordinary nuclei in human-made particle accelerators. These nuclei are often very unstable indeed and break down in mere days, or even in tiny fractions of a second. They can, however, be studied as they break down, and the rules of such breakdowns have been worked out.

When one nucleus breaks down, spontaneously, to another, it is because the second nucleus requires less energy to hold itself together. The nucleus that breaks down gives off energy, in other words, and settles down into a less-energetic combination. That is equivalent to a ball rolling downhill.

A nucleus cannot break down, spontaneously, into another nucleus that contains *more* energy; that would be equivalent to a ball rolling uphill. A ball won't roll uphill unless it is pushed, and a nucleus won't break down to a more energetic form unless energy is pumped into it (and that requires supernovas, cosmic rays, or particle accelerators).

There are various ways in which a nucleus can break down. And when none of these ways will turn a nucleus into a less energetic nucleus, the nucleus simply doesn't break down. It is stable. The 265 stable nuclei I mentioned are those that have no way of breaking down into less energetic combinations.

One common form of breakdown takes place when a nucleus possesses too many neutrons. In that case, a neutron is capable of changing spontaneously into a proton. That often lowers the energy content of the nucleus and produces a stable combination.

A neutron, however, does not carry an electric charge. There are two kinds of electric charge, positive and negative, and you might look upon a neutron as having equal amounts of both, and of therefore being "neutral." (That's how it got its name.)

The proton, naturally, carries a positive electric charge. In order for a neutron to change into a proton, therefore, it has to get rid of the negative portion of any charge it carries. This negative charge is emitted as a speeding electron (also called a "beta particle").

Take the case of a nucleus known as "selenium-82," which is made up of thirty-four protons and forty-eight neutrons. (That's eighty-two particles altogether, which is why it is called selenium-82.) If selenium-82 gives off a beta particle, then one of the neutrons turns into a proton and you end up with thirty-five protons and forty-seven neutrons, which is "bromine-82." However, bromine-82 contains more energy than selenium-82, so selenium-82 can *not* give off a beta particle. It would be rolling uphill if it did; and, since it can't do that, it is stable.

It is bromine-82 that is unstable, for it gives off a beta particle, and one of its neutrons changes to a proton. It ends with thirty-six protons and forty-six neutrons, which is "krypton-82." Krypton-82 has less energy than bromine-82.

The thing is, though, that krypton-82 also has slightly less energy than selenium-82. If selenium-82 gives off *two* beta particles at the same time, it will change to krypton-82. This would be "double-beta decay." This is not likely to happen. Through sheer chance, one particular neutron in the selenium-82 nucleus may be ready to give off a beta particle, but it can't. Or another one might, and *it* can't. Only very rarely are *two* neutrons ready to explode in the same tiny fraction of a second.

Nevertheless, in subatomic physics, as in Nazi Germany, anything that isn't forbidden is compulsory. For a selenium-82 nucleus to change to a krypton-82 nucleus is like going downhill; and, since going downhill is not forbidden, it is compulsory. The selenium-82 nucleus *must* do it by giving off two beta particles, even if very rarely.

Back in 1974, a physicist at the University of California, Michael K. Moe, tested very old rocks that contained selenium-82 atoms. In a billion years or more some of the selenium-82 must have changed into krypton-82, and there must have been tiny traces of krypton-82 present. He looked for the krypton-82 (which is a gas), and he did find tiny traces.

This meant that double-beta decay must take place. Now, it was just a matter of detecting it. If he could gather together in one place as much selenium-82 as possible, he might perhaps catch an occasional nucleus in the act of giving off two electrons.

He collected about half an ounce of it and for years kept trying to catch it in the act of breakdown. The trouble is, all sorts of *other* things kept happening. There are cosmic rays and neutrino bombardment and ordinary radioactivity here and there. There isn't much of any of this, but there's enough to drown out the occasional double-beta decay.

Moe and his coworkers labored to reduce the "noise" and to design detecting devices that would be sensitive only to double-beta decay. Finally, in 1986, he began to detect the double-electron emission. It happened so rarely that half of any quantity of selenium-82 would break down only in almost a hundred billion billion years—the longest "half-life" ever measured. But even so, Moe was dealing with so many trillions of trillions of nuclei that one would break down in this fashion every three or four days. After all, it wasn't forbidden. So it was compulsory.

# 25

# Ozone

Every oxygen molecule is made up of two oxygen atoms. You could call ozone "oxygen and a half," for every ozone molecule is made up of *three* oxygen atoms.

It takes energy to force that third oxygen atom into the combination, and, once the ozone molecule is formed, it is easy for ozone to lose that extra atom per molecule and break down into ordinary oxygen again. In the presence of superfluous energy, as in the neighborhood of electrical-generating machinery, ozone is formed. It doesn't build up to a high concentration, because it tends to break down as quickly as it is formed. This is fortunate, for ozone is poisonous.

Where ozone is chiefly formed is in the higher reaches of the upper atmosphere. There, the energetic rays of the Sun are constantly forming ozone molecules out of oxygen. Again, the ozone tends to break down, but between formation and breakdown an equilibrium is reached, and an ozone layer is built up and maintained in the upper atmosphere, about fifteen miles above the surface.

It isn't really much of a layer, since the air is very thin up there, but it is important—extremely important.

You see, ozone is opaque to ultraviolet light. The ultraviolet portion of sunlight is stopped by the ozone layer and very little of it gets through, while the longer waves of ordinary light penetrate with very little trouble.

This means that when we bask in sunlight, the damaging energetic waves of ultraviolet have been mostly filtered out. What's left is still enough to tan our skins (or burn them, if we are fair-skinned), but by and large we can walk in the sun with reasonable impunity.

On the primordial Earth, there was no oxygen in the atmosphere and nothing out of which to form ozone. It was only the gradual action of the tiny green plants of the ocean, as they evolved, that formed oxygen, which eventually existed in enough quantity to allow an ozone layer to be formed.

It may be that it was not till four hundred million years ago, when the Earth was already over four thousand million years old, that enough oxygen had collected to make it possible to produce a thick-enough ozone layer in the upper atmosphere to shield the Earth. A layer of water absorbed ultraviolet light and protected sea-life, but energetic ultraviolet bathing the bare land would break down the complex chemicals of living things, keeping it sterile. Life could not invade the land, therefore, till the ozone layer had appeared.

But what if something now appeared in the upper atmosphere that would help break down ozone molecules. The present equilibrium would fail, and the ozone layer would get thinner, until it disappeared altogether.

In the early 1970s, two scientists at the University of California suggested that "chlorofluorocarbons" might represent such a danger. These "CFCs" don't burn, aren't poisonous, are absolutely safe to use. They are easily liquefied and then vaporized, so they can be used to transfer heat from one place to another. As a result, after World War II, these chemicals were increasingly used in refrigerators and air-conditioners—and also in spray-cans, where, under pressure, they acted to force the contents out through narrow openings.

Eventually, though, all CFCs leak out of wherever they are and enter the atmosphere. Several million tons have already leaked into the air, and more is added every day. In the atmosphere, they stay. They aren't washed out by the rain or changed by any chemicals. They just drift upward steadily toward the ozone layer.

Once above that layer, the energetic ultraviolet of sunlight is strong enough to break up the CFC molecules and liberate a gas called chlorine. Chlorine, in turn, tends to decompose ozone into oxygen . . . and to thin the ozone layer.

When this was first pointed out, the United States, in the interest of preserving the ozone layer, banned the use of CFCs in spray cans, and other gases were used instead. However, CFCs are still used in spray cans outside the U.S. Furthermore, there are no good substitutes to replace CFCs in refrigerators and air-conditioners.

There were those who argued that the CFCs would not have any serious affect on the ozone layer, but it is possible that they have now been proven wrong.

In 1985, it was discovered that a thin spot, or hole, had appeared in the ozone layer over Antarctica during the autumn. We wouldn't know of this if it weren't for data passed along to us by our satellites. Once the satellite-data, past as well as present, was studied closely, it appeared that the hole had been appearing there for years and had been getting larger year by year. The amount of ozone in the upper atmosphere may have decreased considerably in the last fourteen years, and at this rate it won't be long till the ozone layer is dangerously thin.

This *will* be dangerous. As it is, the ultraviolet radiation that reaches us, despite the ozone layer, is instrumental in producing skin cancer. As more and more ultraviolet penetrates and reaches us, that will increase, especially among those with fair skin. Over the next eighty years or so, there may well be up to forty million cases of skin cancer among Americans alone, and 800,000 deaths. There would also be increases in eye cataracts and other conditions.

If danger to the skin were all, we might just stay indoors as much as possible and carry sunshades out of doors—but it isn't all.

What about other land life?

To protect themselves, higher plants and animals have hair, feathers, scales, cuticles, skin, bark, and so on. Microscopic life-forms in the soil and in the uppermost layers of the sea are not so protected, however. They would find sunshine becoming as deadly as it had been before the ozone layer formed four hundred million years ago. And if these microorganisms are killed, organisms that depend upon them ecologically would surely be adversely affected. In short, it may be that the very fabric of life will be disrupted.

What do we do? Amid all the dangers of overpopulation, of pollution, of drugs, of terrorism, of nuclear war, we must now add our concern for the ozone layer. It is to the ozone layer that leaders in science and government are now increasingly turning their attention.

# 26

# The Ravages of Nature

On September 19 and 20, 1985, twin earthquakes rumbled through Mexico City, the world's largest city. Some 20,000 people were killed, 40,000 more were injured, and 31,000 were left homeless. In that same year, another powerful quake struck in Chile on March 3, leaving 150,000 homeless; luckily, only 177 were killed. Other quakes struck in China and in the Soviet Union.

During the night of November 13, 1985, a long-dormant volcano in Columbia came roaring to life and buried a town at its foot in warm mud, killing 25,000 people who had been sleeping peacefully. Over 60,000 people who survived were either injured or left homeless.

In the northeastern United States, on May 31, 1985, dozens of tornadoes swirled over the land, in the worst such demonstration on the record books. They killed 88 people, injured hundreds, and wiped out whole towns.

Add to this landslides, avalanches, killer typhoons and hurricanes, even a record number of aircraft disasters, and it would seem that nature is on a rampage. Can it be that for some reason our planet has been sufficiently manhandled by us to be shaken out of its equilibrium? Or can it be that certain supernatural powers are annoyed with us?

Will such catastrophes continue to grow continually worse? Or are natural disasters cyclic, and is it that we are just at the bad end of the cycle?

Actually, if we look at such matters carefully, it seems quite clear that natural disasters strike randomly. That does not mean that they do so evenly and that every year is the same as every other. There are both bad years and good years. And if these occur randomly, then every once in a while we may have a long period of relative quiet, and then, unexpectedly, we may have a period in which many years are turbulent indeed. There is (so far) no way of predicting bad years in such a case, or of preventing them.

Yet this suspicion of the randomness of disaster doesn't *seem* right. People my own age can remember years past when disasters happened rarely and were not the everyday occurrences they are now. Why is that?

The answer is that this is an illusion due to the advance of human technology.

Nowadays, our communications satellites, whirling in orbit about the Earth, and the television sets in every home, bring us the details of every disaster at once. For days and days we watched the weary crews striving to rescue one more baby from the piles of rubble into which much of Mexico City had been shaken. We could see the damage, look into frightened, despairing eyes, become totally aware of the scope of the disaster.

Yet it was nowhere near the worst earthquake in history. On January 23, 1556, an earthquake shook down cliffs in northern China, and the collapsing earth buried and killed 830,000 people in the space of five minutes. Europeans at that time never as much as heard of the catastrophe. We know of it only through Chinese records.

Again, on August 27, 1883, the little volcanic island of Krakatoa, between Java and Sumatra, exploded and set off a tsunami (tidal wave) that drowned 36,000 people on nearby coasts. People in Europe and America heard about it afterward with little detail and (of course) with no television pictures. Life went on in the West just as though Krakatoa had never exploded.

But when Mount St. Helens underwent a comparatively tiny eruption in the northwestern United States on May 18, 1980, killing a few dozen people, we watched events nightly on television, saw the climbing columns of smoke and ash, the creeping lava flows, the dust falling on Portland, Oregon, and so on. To Americans, it must have seemed that Mount St. Helens was a much more awesome disaster than the distant, long-ago, hardly noticed Krakatoa.

Then, too, we tend to measure the intensity of a disaster by the number of people killed or the millions of dollars worth of property destroyed, and that, too, tends to make present-day catastrophes seem much worse than those of the past. After all, there are far more people on Earth today than there have been in the past, and they are far more densely packed together and therefore bound to be killed in greater numbers in any catastrophe today than in the case of similar catastrophes a century or more ago.

Again, human structures of all kinds—factories, dams, power stations, high-rise apartments—crowd the Earth as never before, and each one is far more expensive than anything of the sort that has been in the past. Any present-day catastrophe is therefore bound to destroy far more property than such things would have done in the past. For instance, the most dreadful earthquake ever to have struck the United States in its history did *not* take place in California. It took place in the apparently quiet and stable Midwest, as a series of shocks that began on December 6, 1811, and that reached their climax with a ferocious quake on February 7, 1812. The quake was centered near the Mississippi River at the spot where New Madrid, Missouri, now stands.

Trembling was felt as far away as Boston, and 150,000 acres of timberland were destroyed. The course of the Mississippi was changed in several places, new lakes were formed, and some swamps were drained. And yet—we know of not a single person who lost his or her life in this vast disaster. After all, the region was hardly populated at the time. On the other hand, if an earthquake of the same intensity were to take place in the same place *today,* many thousands of people would surely be killed, and many millions of dollars worth of property would surely be destroyed. Even a small earthquake in the same place would nowadays do more damage than the monster quake of 1811–12, and people would surely get the impression that somehow earthquakes were growing worse.

Yet humanity has been spared the greatest disasters of all, disasters that may very well be cyclic. At the present moment, scientists are arguing heatedly over the possibility that, every once in a while, a rain of comets may criss-cross the inner Solar System and that some of these, by sheer chance, may strike the Earth.

A single strike by a comet, with a radius of a few miles, would gouge out a crater a hundred miles wide and throw so much dust into the upper atmosphere that virtually all sunlight would be cut off for weeks or months. A lot of plant life would die, and so would the animal life that depended on it. It is supposed that such a strike, 65,000,000 years ago, wiped out all the dinosaurs, together with many other kinds of plants or animals. Nor was that the worst. About 230,000,000 years ago, over 90 percent of all the kinds of plants and animals then living seem to have been wiped out by such a strike. In each such case (so far) life continued, but it changed drastically, as only the survivors multiplied and replenished the Earth.

Some scientists believe that such a "Great Dying" takes place every 26,000,000 years. The last one seems to have taken place 13,000,000 years ago, so the next one won't take place for another 13,000,000 years. That gives us plenty of time, and perhaps, if human beings or their descendants are still living then, we may have the technology to prevent it.

And meanwhile, we can spot the approach of hurricanes by means of our weather satellites; we are developing methods for predicting earthquakes and volcanic eruptions; and, in general, we are trying to keep from being caught unaware by disasters. That, in itself, should improve matters.

# 27

# The Double Discovery of Evolution

Everybody knows that the English naturalist Charles R. Darwin advanced the theory of evolution, but most people are not quite as familiar with how *long* he worked at it.

The notion of the evolutionary development of life from the simplest creatures to the most complex was in the air in the early 1800s, but no one had advanced a convincing driving force for evolution. What *caused* life-forms to change their characteristics?

Some thought organisms evolved by *trying*. Antelopes that stretched their necks to reach leaves on trees passed longer necks on to their young, and over the generations some antelopes became giraffes. The trouble was, it was easy to show that such acquired characteristics were *not* inherited.

Darwin began thinking of evolution during his voyage on the *HMS Beagle* from 1831 to 1836. His investigations of various life-forms convinced him that evolution took place. But *how,* he wondered.

Then, in 1838, he happened to come across a book written forty years earlier by Thomas Malthus. Malthus maintained that the human population always increased faster than the food supply, so that the population would always have to be cut down by starvation, disease, or war.

It occurred to Darwin at once that this must happen to all species of living things. They *all* outstripped their food supply, and only those who were best adapted to their environment survived. It was this "natural selection" that chose the best adapted in each generation and produced a slow change toward better and better adaptation over millions of years.

Darwin began to collect the evidence he needed. He was in no hurry. He knew that his theory of "evolution by natural selection" would raise a huge storm and that he would be denounced as a blasphemer and an atheist who denied the biblical tale of creation. Darwin was no scrapper, and it was his hope that he would collect so much evidence that no one would be able to argue against him.

He spent years at it. In 1858, twenty years after Malthus had given him his answer, he was still collecting data. He read some of his material to his friends and they urged him to publish, but he was dogged in his pursuit of ever more evidence.

Meanwhile, another English naturalist, Alfred R. Wallace, was also exploring the far regions of the Earth and noting the way in which life-forms differed from each other. He, too, began speculating about evolution. And he, too, wondered what the driving force might be.

Wallace was in Borneo and was immobilized for a time with an attack of malaria. With time hanging heavy on his hands, he happened across a copy of Malthus's book and read it. The same flash concerning natural selection came to Wallace that had come to Darwin.

To be sure, Wallace was far behind Darwin. Wallace read the book in 1858, twenty years later than Darwin had. Wallace, however, was a far more fiery character. Having gotten the key notion, he did not wait. Seizing pen and paper, he began to write, and in *two days* he wrote down his ideas in full detail, with whatever evidence he had happened to have collected in his travels to illustrate his point.

Once his manuscript was done, Wallace wanted to gain the opinion of some other naturalist, and it occurred to him to send it to none other than Charles Darwin.

Darwin received Wallace's manuscript on June 3, 1858, and when he read it he was thunderstruck. His own theory was there before him.

Darwin might have ignored the manuscript and rushed his own material into print. After all, Darwin had far more evidence and a far more carefully reasoned exposition than Wallace had, and many people knew how long Darwin had been working at it. Darwin could easily have gotten away with it.

But Darwin was not that kind of man. He acknowledged receiving Wallace's material and suggested joint publication. An article appeared under both names in a learned journal. The next year, in 1859, Darwin published his researches in full in a book that is usually called *The Origin of Species.* Its entire first printing of 1,250 copies was sold on the first day; and it did, indeed, raise the kind of storm that Dawin had feared. (The storm still rages.) It is this book that gives Darwin the lion's share of credit for the theory of evolution by natural selection.

Unwilling to create more controversy, Darwin did not mention human beings in *The Origin of Species.* However, the controversy came and there was nothing to lose. In 1871, therefore, Darwin published *The Descent of Man,* putting forward all the evidence he could gather for *human* evolution.

Here, however, Wallace balked and would not follow him. It is hard to see how one can accept evolution for all other life-forms, but not for human beings. But Wallace managed that peculiar thing and fell out of the mainstream of biological advance.

In fact, Wallace, as he grew older, adopted a number of odd notions and did so, always, with fiery vigor. He was against Darwin's idea that sexual selection had anything to do with evolution, for instance, and he also took up socialism.

Well, sexual selection wasn't firmly established and socialism was popular at the time among the British intellectuals. Wallace also took up spiritualism, however, something that was harder for scientists to swallow. Most surprising, he vigorously campaigned against vaccination, which was clearly suppressing smallpox at the time. Wallace also took sides in the debate concerning the canals of Mars. In the 1880s and 1890s, a number of astronomers thought they could see those canals and suspected there might be advanced life on Mars. The American astronomer Percival Lowell was quite sure there was life on Mars and wrote several books advocating that point of view.

In 1907, Wallace, then eighty-four years old, was asked to write a review of one of Lowell's books. Boiling with anger and as impulsive as he had been in 1858, Wallace ended by writing a 110-page book of his own, arguing violently against the canals. And in that case Wallace happened to be right, and it was Lowell who was wrong.

# 28

# Master-Lizard, the King

Life developed over three billion years ago, very early in Earth's history. For some 90 percent of that history, however, it existed only in the sea. Then, about 340,000,000 years ago, animal life began to venture out onto land.

At first, it was only a part-way thing. Animals developed ways of living on land and breathing the free oxygen in the air. However, they had to lay their eggs in the water to keep them from drying out. During the early stages of life, the young remained in the sea. Only the adults came out on land. Such animals are called "amphibians," from Greek words meaning "both lives"—that is, both land and water. Today, the most familiar amphibians are the frogs.

But then, about sixty million years afterward, conquest of the land became complete. There were forms of life that laid eggs with porous shells. These allowed air to enter and leave, but they also retained enough water inside to allow the young animal to develop without drying out. Animals with eggs such as these never had to return to the water. They are called "reptiles," from a Latin word meaning "to creep," because the most successful modern-day reptiles are the snakes.

About two hundred million years ago, a group of such reptiles developed that dominated life on land. They spread out over all the Earth's land surface and evolved into many different kinds. They maintained that domination for a long period. For that reason, the period of Earth's history from two hundred million years ago to sixty-five million years ago is called "The Age of the Reptiles." More properly, this period is called the "Mesozoic" era, from Greek words meaning "middle animals." Before two hundred million years ago, fish and amphibians were the most important large animals; after sixty-five million years ago, mammals were. The Age of the Reptiles came in between, in the middle.

Human beings didn't know anything at all about the Mesozoic Era until the early 1800s, when ancient bones began to be found in the Earth. They had been in the ground so long that they had slowly become stony, and they are usually called "fossils" (from a Latin word meaning "to dig up"). Scientists could tell that these bones were from reptiles, even though they were not quite like any reptiles that live today. More bones of this type were searched for and found and gradually, the Age of Reptiles came to be understood in considerable detail.

What impressed scientists (and the general public, too) was that some of the bones were huge. Some of the reptiles of the Mesozoic Era were the largest land animals that have ever lived, for they were much larger than even the largest elephants. In 1842, an English naturalist, Richard Owen, impressed by their size, gave these very ancient reptiles the name of "dinosaurs" from Greek words meaning "terrifying lizards."

The most familiar of the dinosaurs is the "brontosaur," named from Greek words meaning "thunder-lizard," perhaps because the reptile would surely have made a thunderous sound as it walked over the Earth. It had the body and legs of a gigantic elephant. Attached to one end of that body was a long neck with a small head at the end; attached to the other end was a long tail.

A large brontosaur would be sixty feet long from the tip of its head to the tip of its tail, and the highest point of its back would reach to a height of eighteen feet, as high as the tallest giraffe. It might well have weighed as much as thirty-five tons, or three times that of the largest male African elephant.

A similar creature, the "diplodocus" (from Greek words meaning "double beam," which describes the structure of its spine), is more delicately built than a brontosaur, but its comparatively thin neck and tail are much longer. A diplodocus can be up to eighty-seven feet long, even though it may weigh only a third as much as a brontosaur does.

The champion, though, is the "brachiosaur," named from Greek words meaning "arm-lizard" because its forelegs are unusually long, considerably longer, in fact, than the hind legs. It is the largest animal that ever lived on land. It might have been only seventy-five feet long, not as long as a diplodocus, but it would be twenty-one feet high at the shoulders, and it could lift its head forty feet into the air, high enough to peer into a fourth-story window. What's more, it may have weighed as much as eighty tons, more than twice as heavy as a brontosaur.

Naturally, the fossil bones scientists have discovered are very few, and there isn't much chance that we just happened to find traces of the very largest animals, so even eighty tons might not be the true record.

And yet these brontosaurs and brachiosaurs, although their giant size would make them "terrifying-lizards" indeed were we to meet them, are not as dangerous as they might seem. These largest reptiles (like the largest mammals today—elephants, rhinoceroses, and hippopotamus) are plant-eaters. They would not have been interested in us as food, if they were to see us, and they would not have gone out of their way to harm us. Still, if they happened to step on us accidentally, it would very much have felt as though a steamroller ran over us.

Still, not all these giant reptiles were plant-eaters, though the very largest were. Wherever there are large plant-eaters, there are meat-eaters (smaller, but much more ferocious) that prey upon them. Naturally, it would be much worse to encounter such a reptilian meat-eater than a giant plant-eater, just as it would be worse to meet up with a tiger than with a hippopotamus.

Some carnivorous dinosaurs (or "carnosaurs," for short) were larger than others, of course, and in 1902 a fossil-hunter named Barnum Brown discovered bones of a very large carnosaur. Once enough bones had been discovered to build a model of the animal, it proved terrifying indeed. The large carnosaurs proved to be the most ferocious and horrifying animals that ever lived on Earth.

The large carnosaurs stood on two mighty legs, with a thick tail behind. There were also two forelimbs, which were relatively small but could pluck at its prey. The two legs, which alone supported the body, were strengthened by thick dense bones, which were fully needed to support the weight of the mighty beast.

The body, which might be forty feet long and weigh about seven tons, reared up on those two legs to a height of about eighteen feet. It weighed only a tenth as much as a large brachiosaur, but it was much more dangerous. It had a head that was four feet long, and its mouth was fitted with ferocious fangs up to seven or eight inches long.

The large carnosaur discovered by Brown was called a "tyrannosaur," from Greek words meaning "master lizard," and the largest variety was called "Tyrannosaurus Rex" ("master lizard, the king").

It is possible that Tyrannosaurus Rex is not the very largest carnosaur there is. Individual bones, like that of Tyrannosaurus Rex but larger, have been found. Nevertheless, Tyrannosaurus Rex is the largest carnosaur of which we have entire skeletons, and it is fearsome enough to satisfy anybody.

Perhaps the most dramatic effect of the terror such a beast might evoke is to be found in the episode based on Igor Stravinsky's "Rites of Spring" in Walt Disney's motion picture *Fantasia.* No one seeing that film can forget the sudden horrifying appearance of Tyrannosaurus Rex (to the appropriate loud chords of music).

Tyrannosaurus Rex flourished in the last portion of the Mesozoic Era, when there were large plant-eaters for it to prey on. There were, for example, the stegosaurs (one of whom is pictured as fighting the "king" in *Fantasia*), but the very largest—the brontosaurs and brachiosaurs—were long extinct.

This is not surprising. All through the Mesozoic Era, various species of reptiles became extinct, while others evolved, so that the details of the period were always changing. Through almost the entire period there were also tiny mammals, our ancestors, who weren't very successful. This state of affairs might have continued to the very present. There might still be giant reptiles, and mammals might still be small and insignificant, but something happened. . . .

About sixty-five million years ago, very suddenly, all the remaining dinosaurs, as well as other large animals and even many plants and small animals, disappeared. Some small mammals survived, however, and, in the absence of the large reptiles, they prospered and evolved into many large and complex forms—including us.

But what happened to make that possible?

For many years scientists have argued over the matter. In the last few years, evidence has turned up that makes it seem possible that sixty-five million years ago a sizable comet—several miles across—hit the Earth. The impact caused earthquakes and volcanic eruptions, but that was not the worst of it. The cosmic crash threw cubic miles of dust into the upper atmosphere, and there the dust stayed for many months, cutting off the light of the Sun.

In the absence of sunlight, plants died, and then the animals that fed on plants died; and then the animals that fed on animals died. Many forms of life, including Tyrannosaurus Rex, died out to the last individual. Some forms of life, like our mammalian ancestors, managed somehow to survive, feeding on the dead and frozen dinosaurs while waiting for the sunshine to return.

So now, in place of Tyrannosaurus Rex, it is we, *Homo sapiens,* who dominate the world. In many ways, we are more dangerous than Tyrannosaurus Rex ever was.

# 29

# The Hot-Blooded Giants

We tend to think of reptiles as inferior to birds and mammals.

Birds and mammals are "warm-blooded" and can maintain a constant internal temperature, whatever the outside temperature is (within reason). That means they are not sluggish in the cold and need not fear sunstroke in the heat.

Reptiles, on the other hand, are "cold-blooded," and their internal temperature tends to match the outside's. On a cold morning, a lizard is also cold, and its workings are as sluggish as an automobile engine in which the lubricating oil has become thick and sludgy. It has to lie in the sun and warm up as your car's engine does—or as a radio set had to do in the old pre-transistor days—before it can move about energetically. And if the temperature gets too hot and the direct rays of the sun are too harsh, a lizard must seek the shade or it will hard-boil its brain.

The most glorious reptiles that ever lived were the various creatures we know as "dinosaurs," who ruled the Earth until some catastrophe sixty-five million years ago wiped them out. Some were huge beasts that were the most terrifying carnivores who ever lived. Some plant-eaters were even huger, weighing as much as ten elephants and able to reach as high as the roof of a four-story building.

We tend to think of those giants as pathetic monsters, very dim-witted, very clumsy, fit for nothing but extinction. It was inevitable, therefore, that warm-blooded little birds and mammals took over.

But this is nothing more than our own egoism speaking. The dinosaurs ruled the earth for something like a hundred and fifty millions years, and the birds and mammals that came along in the second half of that period were poor specimens indeed. They existed only by hiding out in various nooks and corners and hoping that the dinosaurs wouldn't notice them.

If the catastrophe, whatever it was (very likely a cometary impact and a "nuclear winter"), had not wiped out the dinosaurs, they would probably still rule the Earth, and birds and mammals would still be hunkering down and hiding. The birds and mammals survived the catastrophe that killed the dinosaurs, to be sure, but that had probably more to do with the fact that they were small than that they were quick or brainy.

But if the dinosaurs did so well for so long, might it not be that they were not quite so clumsy, sluggish, and dim-witted? Might they not have been smarter than we think, quicker than we think, *better* than we think? In fact, might they have been warm-blooded?

We can't tell very well, because just about all we have left of them is bones, teeth, and other hard parts. But after all, the mammals evolved from reptiles—not from the dinosaurs, to be sure, but from another and more primitive line of reptiles that are even more ancient than the dinosaurs. These ancestral reptiles are called "theriodonts" ("beast-teeth") because their teeth were more like those of today's mammals than they were like those of other reptiles.

These theriodonts could not compete with the up-and-coming dinosaurs and were extinct by 170,000,000 years ago, but they left descendants that were primitive mammals. Those primitive mammals were surely warm-blooded and had hair. (The hair was needed for insulation, to keep their body heat from leaking away too quickly.) But we don't know exactly when the warm-bloodedness and hair developed. There is no reason not to think that some of the theriodonts developed warm-bloodedness and hair *before* they developed mammalian skeletons and while they were still what we would classify as reptiles.

The birds, which are warm-blooded and which have feathers for insulation, also evolved from reptiles, from dinosaurian reptiles in fact. (There are some paleontologists who say that the dinosaurs never did become extinct. They just sit in the trees and sing.) Possibly, here again it may be that the forebears of the birds developed warm-bloodedness and feathers before they became entirely birds. In fact, the "archeopteryx," the earliest known organism with feathers, had a lizard-like head, lizard-like teeth, and a lizard-like tail.

If two different groups of reptiles developed warm-bloodedness, perhaps it was more widespread in the group than we think. Perhaps it is only an accident that the few reptiles that survived the catastrophe happened to be cold-blooded.

There are paleontologists who feel that the dinosaurs could not have been as successful as they were if they were not warm-blooded; that at least some of them had to be quick and active; that they were *hot-blooded* giants, in other words. The arguments usually rest on some pretty tortuous reasoning, and the arguments among palentologists on the matter are even warmer than the dinosaur blood may have been.

There are, however, dinosaurian relics other than petrified skeletons. Back in the 1920s, the American paleontologist Roy Chapman Andrews discovered fossilized dinosaur eggs in central Asia. This somehow humanized dinosaurs, since we could think of the creatures as parents.

And now paleontologists have discovered an enormous cache of dinosaur eggs in the southwestern corner of the Canadian province of Alberta, near the American border. There are nests upon nests, each with a number of eggs in it. Apparently, this was a nesting ground for a kind of dinosaur called "hadrosaurs." But the site was drowned out when a river flooded the area about seventy-three million years ago.

The mere fact that the nests are so many leads one to think of the hadrosaurs as being almost mammalian or bird-like in their care of their young. In turn, this makes them seem more "advanced" and makes warm-bloodedness easier to accept. (Of course, alligators take good care of their eggs and young, and they are the closest living relatives of the dinosaurs—and they are cold-blooded.)

Some of the eggs have advanced fetuses inside, fossilized, of course, and they can be studied carefully.

The presence of all those eggs makes it possible to measure the rate at which the fetus, and for that matter the hatched-out young had grown. The faster the growth-rate, the more likely the creatures were warm-blooded. Preliminary results seem to encourage the warm-blooded view. But, as yet, there are a great many more eggs and fetal remains to study.

# Part IV
# SciQuest

Beginning in 1980, I undertook to write a column for a magazine called *SciQuest*, which was published by the American Chemical Society. It was intended for high-school chemistry students, and my essays, five-hundred words a piece, dealt for the most part with scientists (not necessarily chemists) and their triumphs and troubles.

I enjoyed doing the column very much, and I kept it up until (alas!) the American Chemical Society decided that the magazine wasn't attracting sufficient readership. They therefore brought it to an end, and naturally my column came to an end, too.

I have here a selection of twenty of the essays I wrote for *Sciquest*. It is not all that I wrote, of course, but I felt that twenty of them would be enough of a sampling.

In connection with the essay "Sometimes it Takes Time," which describes how Francis Peyton Rous had to wait fifty-five years to get a Nobel Prize for his discovery, I would like to add a very similar case that took place since that essay was published. In 1932, the German electrical engineer Ernst A. P. Ruska invented the electron microscope. He received a share of the Nobel Prize for physics in 1986 for this invention—fifty-four years later.

# 30
# The Absent-Minded Professor

One long-standing bit of folklore is that professors are absent-minded. Partly, this arises out of the notion that they must be concentrating so hard on very difficult matters that they fail to notice the ordinary, everyday things going on all about them.

Perhaps the oldest absent-minded professor story in literature is one about Thales, a Greek philosopher (624-546 B.C.) who lived twenty-five centuries ago. Another philosopher, Plato (427-347 B.C.), writing two centuries later, tells of Thales walking along one night, intently studying the stars. Not noticing where his feet were taking him, Thales fell into a well. An old woman coming in response to his cries helped him out, then said with contempt, "Here is a man who would study the stars and cannot see what lies at his feet."

Like almost all absent-minded professor jokes, this one seems apocryphal. It may have been made up by Plato just to point to a moral.

Another Greek philosopher, Archimedes (287-212 B.C.), discovered the principle of displacement while he was at the public baths. Excited beyond endurance, he completely forgot that he had no clothes on and rushed home naked through the streets of the city, shouting, "Eureka! Eureka!" ("I've got it! I've got it!"). This may very well have been true, but it might not have been so terribly absent-minded of Archimedes after all. The ancient Greeks weren't bothered as much by public nudity as we are.

Suppose we come closer to the present. There is the case of Karl F. Gauss (1777-1855), a German mathematician who, many people think, was the greatest mathematician who ever lived. He made important discoveries while he was still in his teens, and continued to grind out important papers to the end of his life. Naturally, he was always engrossed in one deep problem or another.

There is a story that in 1807 Gauss's wife was very ill and the doctor was in the bedroom attending her. Gauss, sad and miserable, was waiting below, when his eye caught the papers on which he had been working out some problem. Little by little, his thoughts turned to it again, and very soon he was entirely occupied with it. And while he worked away, the doctor came down to tell Gauss the sad news that his wife was dying.

Gauss, his mind still firmly fixed on the problem, waved absently at the doctor and muttered, "Yes, yes, but tell her to wait a moment till I'm through."

A great number of stories are told of the American mathematician, Norbert Wiener (1894-1964). One of them describes a time he met an M.I.T. colleague on Memorial Drive. They both stopped, began to talk, and went on to discuss many things. After half an hour, it was clear that they were through and were going to separate.

They had shaken hands when Wiener hesitated and said, "By the way, when we met, was I walking toward Massachusetts Avenue or away from it?"

His colleague said, "Why, away from it, Norbert."

"Good!" said Wiener. "Then I've had my lunch!"

I knew Dr. Wiener and I would believe the story in a minute, except that they tell it about other professors, too.

# 31

# Playing It Safe

Sometimes a scientist can be caught between the desire to get credit for a major discovery and the desire not to look foolish. In 1931, Fred Allison, of the University of Alabama, reported the isolation of elements 85 and 87 and named them "alabamine" and "virginium," respectively.

But he was wrong. It wasn't for another decade that they were really discovered—and named "astatine" and "francium." Allison is now remembered chiefly for his error.

This is not something new. Back in 1610, Galileo had to be very careful. His important new discoveries were derided by some who insisted that his telescope could not be trusted and that the new objects he discovered in the heavens were illusions created by the lenses. Galileo was intent, therefore, on holding back his findings until repeated observations under different conditions convinced him they could not be illusions.

On the other hand, others were likely to be hot on his heels and claim precedence. Within a couple of years, in fact, the astronomer Simon Marius claimed to have discovered the four large satellites of Jupiter before Galileo did, while another astronomer, Christoph Scheiner, claimed to have observed sunspots before Galileo did.

Toward the end of 1610, Galileo discovered that Venus showed phases just as the Moon did, going from new-Venus to half-Venus to full-Venus, then back to half-Venus and new-Venus. This was important: By the old Ptolemaic theory Venus could not show phases in this fashion; by the new Copernican theory, it had to.

The importance of the discovery was therefore first-rate. It would be the last nail in the coffin of Greek astronomy and would finally establish the Sun as the center of the planetary system. On the other hand, Galileo's telescope could just barely make out those phases, and if he was being misled by wishful thinking, it might cast a pall over all his other discoveries.

What Galileo did, therefore, was to play it safe. On December 11, 1610, he sent a letter to a friend of his, Giuliano de Medici, an ambassador in Prague. It contained the following sentence in Latin: "Haec immatura a me jam frustra leguntur o.y." This means, "These unripe things are read by me." The sentence hints Galileo has made a discovery but that he isn't ready to say what it is.

The final "o.y." are two letters left over, which show the whole message to be an anagram. When the letters are rearranged with the "o" and "y" included, a different message can be obtained.

If the phases of Venus proved to be a false alarm, the message to de Medici would remain as it was and mean nothing. If, however, Galileo grew convinced of the actual existence of the phases, he could present the rearranged message as "Cynthia figuras aemulatur Mater Amorum." This means, "The Mother of Love imitates the shape of Cynthia," where "the Mother of Love" is Venus, of course, and "Cynthia" is a poetic name for the Moon.

Furthermore, if anyone else announced the discovery of the phases of Venus while Galileo was waiting, Galileo could unscramble the anagram at once and use the date of the letter and the word of a respected ambassador to establish his own precedence.

There was a happy ending. Galileo's observations were correct and he got the full credit without dispute.

# 32

# The First Scientist

Who was the first scientist?

It wasn't Isaac Newton. Today, it is generally acknowledged that Newton was not only a scientist, but the greatest scientist who ever lived, yet Newton never thought of himself as a scientist. He couldn't, for the word didn't exist in his time.

Newton thought of himself as a "philosopher," a word that dates back to the ancient Greek thinkers and that comes from Greek words meaning "lover of wisdom."

There are different kinds of wisdom we might love, of course. Some philosophers are concerned chiefly with the wisdom derived from the study of the world about us and the manner of its workings. The world about us can be referred to as "nature," from the Latin word meaning "birth." Nature, in other words, is everything that has been created or that has come into being. Philosophers who deal primarily with nature are, therefore, "natural philosophers."

Newton thought of himself as a natural philosopher, and the sort of thing he studied was natural philosophy. Thus, when he wrote the book in which he carefully described his three laws of motion and his theory of universal gravitation—the greatest scientific book ever written—he called it (in Latin) *Philosophiae Naturalis Principia Mathematica* which in English is *The Mathematical Principles of Natural Philosophy.*

The Greek word for "natural" is *physikos,* which in English becomes physical. Natural philosophy might also be spoken of as "physical philosophy," which can be shorted to "physics."

As natural philosophy grew and expanded, all kinds of special studies developed. People began to speak of chemistry, of geology, of physiology, and so on. Physics was whatever was left over, so it didn't suit as a general overall word for natural philosophy. Yet you needed some such short word, for natural philosophy was a seven-syllable mouthful.

There did, for instance, exists the word "science," from the Latin word meaning "to know." Originally that meant knowing about anything at all, so that if you knew how to play a basketball game, it was fair to say that you understood the science of basketball.

Gradually, though, because something was needed that was short and convenient as a word expressing the kind of knowledge that natural philosophers were interested in, "science" came to be used to mean "natural philosophy."

Then, about 1840, an English natural philosopher named William Whewell began to use the word "scientist" to represent someone who studied and understood that kind of science. In other words a scientist came to mean a natural philosopher.

Only after 1840, then, can there be anyone who thinks of himself as a "scientist." In that case, who was the first scientist?

Well, Whewell was a good friend of Michael Faraday and suggested a number of new words for concepts that Faraday had dreamed up, words such as "ion", "anode," "cathode," and so on. What's more, Faraday was the greatest natural philosopher of his time, one of the ten best of all time surely, and probably the greatest experimenter of all time.

If Whewell thought of anyone as a scientist, I'll bet he thought of Faraday first. And if he didn't, I will.

I say Michael Faraday was the first scientist. And the first physicist, for that matter, since Whewell also made up that word.

# 33

# Tough Luck

We've all had our share of bad breaks, but some of us do seem to get it in the neck a little more than others do. Consider the case of Guillaume Le Gentil, a French astronomer who in 1761 wanted to observe the transit of Venus.

In those days it was felt that if accurate observations were made from widely different positions on Earth of the exact moment Venus moved in front of the Sun's edge, its exact distance from the Sun could be calculated. The method turned out to be defective because Venus's atmosphere made the exact time of apparent contact with the edge rather fuzzy. But in 1761 this was not known, and Le Gentil was extremely anxious to try.

What's more, he was going to go to Pondichéry, India, a French-controlled settlement, in order to get a measurement far removed from the measurements to be made in Europe.

France and Great Britain were at war, however, and just as Le Gentil arrived at Pondichéry, that city was taken by the British, who wouldn't allow the French ship to dock.

Le Gentil tried to witness the transit from on board the ship, but the ship pitched and rolled, and there was no possibility of a decent measurement. His long, uncomfortable trip from France had come to nothing.

As it happens, transits of Venus come in pairs, the second one coming eight years after the first. It is then necessary to wait over a century for another pair. The transit of 1761 was the first of a pair, and there would be another one in 1769.

Le Gentil might take a ship back to France and spend many weary months on the stormy seas. But then he would have to take another ship and spend many weary months coming to India a second time. He decided he couldn't face that, so he made up his mind to remain in India and wait eight years for the second transit.

He didn't waste his time, and in the course of the eight years he learned all he could about India and made many observations of weather, tides, ancient Indian astronomy, and so on. He also traveled about southern Asia. He reached Manila in the Philippines and for a while wondered if he ought to make the measurement of the transit there, for his calcuations showed it would be splendidly visible from Manila.

However, all the calculations in Europe were made with a Pondichéry observation in mind and Le Gentil decided to go back to Pondichéry (which the British had long since returned to the French) and be ready on June 3, 1769.

There he was with his instruments in place. In Manila, as it turned out, the weather was perfectly clear all day. In Pondichéry, it was clear, too—before the transit . . . and after the transit. At the exact time of the transit, however, a dark cloud moved in front of the Sun, and the eight-year wait turned out to be completely in vain. Le Gentil had missed again.

Miserable and unhappy, he returned to France in 1771, after an absence of eleven and a half years.

His bad luck wasn't over, though. Somehow his messages to France had never arrived, and his relatives had decided he was dead and had divided his property among themselves. Le Gentil managed to legally establish that he was alive, but he couldn't get his money back and he had to pay for his court costs.

But all was not lost. Le Gentil started life anew. He married, had a daughter, wrote a two-volume book on India, and lived on, quite successfully, for twenty-one years.

And after all, even if he *had* observed the transits, they would not have yielded the results he expected because of the existence of Venus's atmosphere.

# 34

# To See Is Not Enough

On the night of March 13, 1781, an amateur astronomer named William Herschel was meticulously scanning the heavens. He came across a body that showed a visible disc. He thought he had discovered a new comet. He kept an eye on it and found it moved against the background of the skies more slowly than a comet usually did. What's more, it showed a sharp boundary rather than a comet's hazy one. Eventually, he decided he had made an unprecedented discovery—a new planet. It was the first planet discovered in modern times, the one we now call Uranus.

Uranus is an object of the sixth magnitude, which means it is just visible to normal eyes on a clear, Moonless night. People couldn't be blamed for not noticing it when it was so dim and, being so distant from the Sun, moved so much more slowly against the background of the stars than the other planets did. At least there was no blame if the observations were being made with the unaided eye.

With a telescope, however, it ought to have been easily visible. How was it that no one had noticed Uranus in the nearly two centuries that telescopes had existed before Herschel? It was impossible to suppose so. It *had* been noticed.

In 1690, nearly a century before the discovery of Uranus, John Flamsteed, England's astronomer-royal, while mapping the heavens observed a sixth-magnitude star in the constellation of Taurus, and he carefully recorded its position. Using his system for naming the stars, he called it "34 Tauri."

No one else ever observed that star in that position—because it wasn't a star. It was Uranus, and it moved steadily out of that position. How do we know? Because after Uranus was discovered its orbit was calculated and older star charts were studied to see if any star was reported where no star existed but where Uranus would have been at that time.

In fact, Flamsteed recorded Uranus on four additional occasions in four different positions in its orbit.

Flamsteed died in 1719 and never knew what he had missed, but then he was not the only one to observe Uranus before it had been discovered. There were, in fact, astronomers alive at the time of the discovery of Uranus who wondered whether they had made similar mistakes.

One was Pierre Charles Lemonnier, who was sixty-five years old at the time of Herschel's discovery. He went over his own records and found that he had seen Uranus on three different occasions and recorded its position in three different places.

Another astronomer, Alexis Bouvard, also went through Lemonnier's records and found not only those three observations but ten others, four of them on consecutive nights! Lemonnier couldn't possibly have missed discovering the planet, if he had only compared his own charts.

Ought we to sneer at those astronomers who missed their opportunity? Of course not! A telescope is not a telescope; some are better than others.

Herchel, a dedicated amateur, built his own telescopes; better ones than any that existed at the time. His was the first that could make out Uranus as a tiny disc and not just as a dot of light. It was the disc, not the motion, that attracted his eye. Otherwise he might have missed it, too.

# 35
# The Race For Honor

Scientists are human. Unraveling the knots of Nature's mysteries is a reward in itself; but even so, scientists like to hear the applause of the audience.

In the world of science, the first person who publishes gets the credit for a discovery, not necessarily the first person who actually makes the discovery. Nothing is of importance to the world of science till it is made known to scientists generally.

In the 1830s three scientists were independently attempting to measure the parallax of a star in order to make a first accurate estimate of stellar distances. The first to complete the observations and calculations was Thomas Henderson, who had been observing Alpha Centauri from the Cape of Good Hope. Second was Friedrich W. Bessel, who had been observing 61 Cygni from Königsberg.

However, Henderson was going home and decided to wait till he was back in Scotland before preparing a paper on his work. Bessel was already home and didn't have to wait. Bessel published in 1838, Henderson in 1839. The result? Look in any astronomy textbook and you'll find that the first man to measure the parallax of a star and to determine its distance was Bessel.

Sometimes, it isn't even the loser's fault. In 1771 and 1772, Karl Wilhelm Scheele isolated oxygen from a variety of substances, including mercuric oxide. He observed its properties and wrote up his findings in publishable form.

In 1774, Joseph Priestley isolated oxygen from mercuric oxide, observed its properties, and wrote up his findings in publishable form.

Scheele was clearly first, but the publisher to whom he had entrusted his paper was negligent beyond belief and didn't get around to publishing the paper till 1777. By that time, Priestley's paper had already been published. The result? Look in any chemistry textbook and you'll find that Priestley was the first to isolate oxygen.

Sometimes even publishing doesn't help. In the 1740s, Mikhail V. Lomonosov published papers arguing against the phlogiston theory of combustion and suggesting that mass was conserved in chemical change.

In the 1770s, Antoine L. Lavoisier published papers arguing against the phlogiston theory of combustion and suggesting that mass was conserved in chemical change.

Lavoisier's work *was* the more systematic of the two and the more convincing, but Lomonosov beat him by thirty years and yet gets no credit at all. Any chemistry textbook will tell you that it was Lavoisier who destroyed the phlogiston theory and established the conservation of mass. Lomonosov is likely not even to be mentioned.

Why? Lavoisier published in his native French; Lomonosov in his native Russian. At that time, all scientists could read French or have it translated, but no scientist outside Russia (where there were very few scientists, then) dreamed of reading Russian.

There is injustice in the best-regulated society—even in the world of science.

# 36
# Thoughts In Prison

We have all heard of literary men who have composed masterpieces while in prison. John Bunyan wrote *Pilgrim's Progress* while imprisoned in 1675, for instance. Well, there have been occasions when important scientific advances have taken place in prison. For instance. . . .

Jean Victor Poncelet graduated as a military engineer in 1810. He was with the French army as a lieutenant when Napoleon invaded Russia in 1812. The invasion was a disaster for Napoleon. At the Battle of Krasnoye, fought on November 16, 1812, Poncelet stopped a bullet during the French retreat and was left for dead on the battlefield.

A group of advancing Russians noted that one of the "corpses" was feebly moving. They took Poncelet to shelter and treated his wounds. When he could walk, he was marched, with other prisoners, a thousand kilometers eastward to Saratov, on the Volga River, a march that took four months of a Russian winter. Poncelet survived that, too. In Saratov, Poncelet remained in prison for a year and half and survived even that. He finally returned to France in 1814, after Napoleon had abdicated.

What helped him to survive in prison was his meditation on geometry. He imagined what would happen if geometric figures cast shadows and how those shadows would change as the geometric figures turned and tilted. There would be strict rules governing those changes, and Poncelet could see that geometric problems that were hard to solve in the ordinary way could be easily handled by this "projective geometry."

In 1822, he published a book on the projective geometry he had thought about in prison, and that book is often viewed as having founded modern geometry.

Next take the case of a French geologist, Dieudonné de Dolomieu. He was the son of an aristocrat who enrolled him in the prestigious Order of the Knights of Malta in 1752, when he was two years old. By the time he was thirty, he had risen to the rank of Commander, but he was a tactless person who had made many bitter enemies among the other Knights. Dolomieu was also interested in science, and the study of minerals was his hobby. He had an excellent mineralogical collection.

In 1798, he accompanied the young Napoleon Bonaparte on the latter's invasion of Egypt. When Dolomieu was returning to France, his ship encountered a storm and was forced into harbor at Taranto, in southern Italy. The region was at war with France then, so Dolomieu was treated as a prisoner (something his enemies in the Knights of Malta made sure of) and was kept in solitary confinement for nearly two years.

Solitary confinement can lead to mental breakdown if a prisoner doesn't work hard to remain stable. Dolomieu managed by concentrating very hard on his mineralogical collection and thinking about the general rules that seemed to govern their appearances and properties.

He carved himself a pen out of wood, and he mixed soot from his lamp with water in order to make a kind of ink. He then used as a writing suface the only paper he had—his Bible. Carefully, as the months passed, he wrote down his thoughts and his remembered observations in the margins of that Bible.

When he was finally released in 1800, those marginal notes formed the basis for two important books on mineralogy, which he published in 1801.

# 37

# Getting Started

The usual way to get started in a career in science is to go to school, take a variety of complicated courses, and gain a number of degrees—as I did.

In past generations, however, it wasn't that easy and some of the greatest scientists made their start in other ways. Michael Faraday, for instance, born in 1791 in England, was one of ten children of a blacksmith. There was no question of any education for him past reading and writing. When he was fourteen, he was apprenticed to a bookbinder.

Fortunately, his master allowed him to read some of the books that were being bound, and in this way Faraday began to teach himself electricity and chemistry. When he was twenty years old, a customer gave him tickets to attend the popular science lectures of the great chemist Humphry Davy. Young Faraday took careful notes, added colored diagrams, and ended with 386 pages, which he bound in leather.

He sent the result to Davy and asked for a job as an assistant. Davy was impressed and hired Faraday as a bottle-washer, at a lower salary than he had earned as a bookbinder. Faraday was treated as a servant at first, but little by little his abilities shone through and, after a dozen years had passed, it became clear that Faraday was to be a greater scientist than Davy. (Davy never forgave him for that.)

Joseph Henry was born in 1797 in Albany, New York. Like Faraday, he was born into a poor family and had little schooling. When he was thirteen, Henry was apprenticed to a watchmaker, so he didn't even have Faraday's good fortune of finding books at hand.

When he was sixteen, Henry was on vacation at a relative's farm. One day he tried to catch a rabbit, which fled under an old church building. Determined to get the rabbit, Henry crawled underneath the church building, too, and found some of the floorboards missing. That caused him to lose interest in the rabbit, since it seemed more interesting to explore the church.

Inside the church, Henry found a shelf of books. One book was called *Lectures on Experimental Philosophy,* which dealt with the new discoveries of science. Henry began leafing through it, stopped to read, and was inspired with curiosity and ambition. The owner of the book let the young man keep it, and Henry returned to school.

He entered the Albany Academy and taught himself rapidly by outside reading as well. In order to pay the tuition and to earn his keep, he engaged in private tutoring and in teaching at country schools, passing on the knowledge he was gaining for himself.

In the 1820s and thereafter, he and Faraday, working independently on opposite sides of the ocean, made world-shaking discoveries in electricity. Faraday invented the electric transformer and the electric generator. Henry invented the electromagnet and the electric motor. Together, they electrified the world.

Faraday became so famous that he was invited to dinner by Queen Victoria; and when Henry died, President Rutherford B. Hayes attended the funeral. Not bad for poor boys who began life with no advantages at all—except brains, ambition, and industry.

# 38
# The Moon Hoax

Many scientists wonder why the general public falls for such patent nonsense as Velikovsky's theories about Venus, or von Däniken's speculations about ancient astronauts, or Berlitz's tales about the Bermuda Triangle. But why not? Nonsense has always found a ready audience.

Back in 1835, the *New York Sun* was founded. The new newspaper needed to grab the public's interest if it were to survive, so the editors looked about for a writer who could turn out interesting copy. They came upon Richard Adams Locke, who had arrived from England only three years before, and they hired him to write essays for the paper.

Locke had tried his hand at writing science fiction, and now it occurred to him that he could write some more without actually saying that that was what it was.

For his subject, he chose the expedition of the English astronomer John Herschel, who had gone to Cape Town in southern Africa in order to study the southern sky. Herschel had taken good telescopes with him, but they were not the best in the world. They didn't have to be. Their value lay not in themselves but in the fact that since all astronomers and observatories were at that time located far to the north of the Equator, the regions near the South Celestial Pole had scarcely been studied. Almost any telescope would have been useful.

Locke improved on that. Beginning with the August 25, 1835 issue of the *Sun,* he carefully described the nature of Herschel's expedition. Locke wrote that Herschel's telescope was so powerful that it could make out objects on the Moon's surface that were as little as eighteen inches across.

In succeeding installments, the surface of the Moon, as seen through this marvelous telescope, was described. Herschel, Locke said, had seen flowers like poppies, and trees like yews and firs. A large lake, with blue water and foaming waves, was described, as were large animals resembling bison and unicorns.

One clever note was the description of a fleshy flap across the forehead of the bisonlike creatures, a flap that could be raised or lowered to protect the animal from the extremes of light and darkness.

Finally, creatures with human appearance, except for the possession of wings, were described. They seemed to be engaged in conversation. "Their gesticulation, more particularly the varied action of their hands and arms, appeared impassioned and emphatic. We hence inferred that they were rational beings."

(It's a shame that Locke didn't think of equipping Herschel with miraculous earphones, for he could have described the astonomer, then, as being able to hear what the Moon-creatures were saying.)

Astronomers recognized the story to be nonsense, of course, since no telescope then built (or now, either) could see such detail from the surface of the Earth. To be sure, what Locke described was completely at odds with what was known of the airlessness and waterlessness of the Moon.

Locke was forced to admit it was all a hoax soon enough. And when Herschel came back and learned about the hoax, he just laughed.

However, the *Sun* got exactly what it wanted. It was now a lasting success. In fact, while Locke's essays were being printed, its circulation soared until, for a brief moment, it became the best-selling newspaper in the world.

The moral is that nonsense is always more easily believed than sense is. I'll leave it to each of you to decide why that should be.

# 39

# Scientific Heretics

Think of a scientific heretic as someone who advances observations, conclusions, or theories that oppose a generally accepted scientific belief, who is persecuted for it, and who nevertheless is found to be right in the end. There turn out to be surprisingly few.

Almost all startling advances in science must step on the toes of earlier beliefs, and conservative scientists do not promptly fall behind the new ideas. Lavoisier's theory of combustion, Dalton's atomic theory, Joule's notions on the conservation of energy, Mendeléev's periodic table, Planck's quantum theory, Rutherford's nuclear atom, and Einstein's relativity were all greeted with hesitancy and doubt, and were denounced unsparingly by conservatives.

By and large, however, these breakthrough scientists did not suffer for their temerity. Their careers proceeded onward triumphantly; they were supported by many scientists, particularly the younger ones; and they received rewards and high regards in their own lifetimes.

There are, of course, cases of the reverse. In 1836 the French chemist

Auguste Laurent advanced a new theory of molecular structure that went against the ideas of the aged demigod of chemistry, Berzelius. The old man denounced Laurent's notions so powerfully that Laurent's career was ruined. Since Laurent died in his forties, he did not live to see his ideas win acceptance.

Again, the German geologist Alfred Wegener suggested in 1912 that the continents drifted slowly and that they all formed a single body of land some hundreds of millions of years ago. He was laughed out of court. Since he died at fifty, he didn't live to see his idea of continental drift (greatly modified) win acceptance at last.

Then, in 1911, the American physician Francis P. Rous advanced the first evidence for the existence of a cancer virus, when such viruses were not acceptable to established thought. Rous did not win the Nobel Prize the finding deserved, therefore, till 1966—fifty-five years after the discovery. Fortunately, he was still alive and at the age of eighty-seven accepted the prize for an idea that time had made respectable at last.

The scientific heretics who *really* suffered denunciation and misery were those whose beliefs threatened *not* older scientific notions, but dogmas *outside* science. Then the forces of religion and popular emotion rose against them.

When Copernicus and Galileo advanced ideas that threatened the motionless, central Earth of the Bible, when Darwin's evolutionary suggestions threatened the special creation of Man, or when Hutton and Lyell presented evidence to refute the notion that Earth was created 6,000 years ago, the people raged. Copernicus dared not publish until he was dying; Galileo was threatened with torture; and the others underwent vilification by a public that would have killed them if it could.

On the other hand, heresies that cater to popular superstition are greeted with enthusiasm. Let someone try to explain biblical miracles with half-baked astronomy, or talk about flying saucers filled with the rough equivalent of angels or demons, and he or she is enshrined by an admiring public who promptly compare these nonsense-mongers with Galileo.

If they *really* resembled Galileo, of course, the public would try to tear them apart.

# 40

# Gold From The Sun

There are some people who aren't impressed by scientific discoveries, as witness the following tale.

The German physicist Gustav Robert Kirchhoff, working with the spectroscope and the Bunsen burner, had discovered by 1859 that each element seemed to produce a characteristic pattern of spectral lines when heated to incandescence. In a way, he had worked out a system of "fingerprinting" the various elements.

If a mineral is heated to incandescence, and if spectral lines appear that do not duplicate the position of those of any known element, the conclusion must be that an unknown element is present.

In 1860, Kirchhoff heated a certain mineral and detected a blue line he could not identify. Using that as a guide, he tracked down a new element, which he named "cesium" (from the Latin word for "sky-blue"). In 1861, he tracked down a red line to discover "rubidium" (from a Latin word for "red").

Kirchhoff went further. He noticed that the bright double-line of yellow in the sodium spectrum was in exactly the same position as a dark double line in the Solar spectrum. He wondered if, when light passed through the cold gas, that gas would absorb just those spectral lines it would emit if it were incandescent. He checked this by experiment, and the rule is called "Kirchhoff's Law."

The dark sodium line in the Solar spectrum could be explained by supposing that light from the hot surface of the Sun passed through sodium vapor in the Sun's atmosphere, which was somewhat cooler than the glowing surface. In this way, Kirchhoff showed that sodium existed on the Sun, and half a dozen other elements as well.

This was a very dramatic discovery. In 1835, the French philosopher Auguste Comte, speaking of the limitations of human knowledge, claimed that no one could possibly discover the chemical makeup of the stars. Well, human knowledge does have its limitations, but Comte's example proved a bad one.

But, as I said, some people are not impressed by such things. Kirchhoff's banker, a highly practical man, said, with an offhand wave of his hand, "Suppose you do discover gold in the Sun. Of what use is that gold if it cannot be brought down to Earth?"

Eventually, when Kirchhoff was awarded a medal and a prize in golden sovereigns from Great Britain for his spectroscopic work, he deposited the sovereigns with his banker, saying wryly, "Here is gold from the Sun."

Spectroscopy went on to enlighten science in even more startling ways. It made it possible not only to determine the chemical makeup of stars, but their temperatures as well. This, in turn, made it possible to classify stars into groups and to allow human beings to learn how stars evolved.

The spectral lines could tell us the rates at which stars and other heavenly objects moved toward or away from us, and that made it possible eventually to show that the Universe was expanding and to measure the distance of objects billions of light-years away. Spectral lines have also given us insights into the intimate structure of the atom.

Bankers may shrug their shoulders at this, but to scientists such knowledge is worth infinitely more than gold.

# 41

# The Joys Of The Unexpected

As might be true of anyone else, a scientist can be engaged for years in work that, while useful, is rather predictable and run-of-the-mill.

As is true of almost no one else, however, a scientist can run into the unexpected at any moment and find him- or herself the agent for transforming society, or the human view of the Universe.

In 1887, for instance, A. A. Michelson and E. W. Morley had devised a delicate "interferometer," which would help them determine the direction and speed of motion of Earth through the basic fabric of the Universe. Everyone knew the Earth was moving; it was just a matter of putting a figure to it. Michelson and Morley, however, to their astonishment, failed utterly. Earth, it seemed, was not moving at all.

That failure, however, led to a new look at the Universe, a questioning as to whether it had a basic fabric at all, and finally, to Einstein's theory of relativity. That's what can happen to an experiment that fails.

In 1883, Thomas Edison was doing his best to work out some way of making the filaments in his newly invented electric light bulb last longer. One of the things he tried was to seal a metal wire into a light bulb near the hot filament, just to see if that would help. Edison noted that electricity flowed from the hot filament to the metal wire across the vacuum between them, but that didn't seem to help make the filament last longer, so he thought no further about it. (He recorded it and patented it, however.)

It was this "Edison effect," in the hands of others, that led to the birth of the vast electronics industry. Modern radio, television, and numerous other devices grew out of that apparently useless finding.

In 1927, Clinton Davisson was studying the reflection of electrons from a metallic nickel target enclosed in a vacuum tube. He expected to get some interesting but entirely undramatic data—but then the tube shattered by accident, and the heated nickel promptly developed a film of oxide that made it useless as a target. To remove the film, Davisson had to heat the nickel for a long time.

Without his knowing it, that changed the nickel surface from many tiny crystals to a few large ones. When he used the new surface for electron-reflection, he found, unexpectedly, that the electrons behaved as though they consisted of waves. It was only because he had large crystals on the surface that the effect (predicted but never observed) had showed up. Eventually, he received a Nobel Prize for the discovery, which might not have come his way but for a laboratory accident.

In 1967, Anthony Hewish devised a new radio telescope designed to detect very rapid radio-wave fluctuations, hoping to add a few more possibly interesting details to what was already known. Using the telescope, his assistant, Jocelyn Bell, unexpectedly discovered very rapid, very regular bursts of radiation that proved to come from a type of object hitherto-unknown: a pulsar or neutron star—something with the full mass of an ordinary star, but no more than a few miles across. Eventually Hewish got a Nobel Prize for that.

That's science for you. No matter how dull a particular job may seem, there could be something world-shaking waiting for you right around the corner.

# 42

# Facing the Giant

William Thomson, a giant of nineteenth-century science, was in his old age when he was given a title and became Lord Kelvin. But he was only in his twenties when he calculated how old the Earth must be.

Suppose, he said, the Earth had once been part of the Sun and had been thrown off (which was thought likely at that time). We know how hot the outer layers of the Sun are; we know how cool the outer layers of the Earth are. How long would it take the Earth to cool down from the Sun's temperature to its own present temperature?

Answer: from 20 to 400 million years.

Later, Kelvin considered the Sun itself. He accepted the then-current theory that its radiant energy was derived from the gravitational energy made available as it slowly shrank. How long would it take for it to shrink from a size large enough to include Earth's orbit to its present size—if it did so at a rate fast enough to supply the energy it radiated away?

Answer: about 25 million years.

So it seemed, by unassailable physics and mathematics, that the Earth could only be a couple of dozen million years old. This horrified geologists, who were certain the Earth had to be ten to a hundred times as old as that. But these geologists lacked the arguments to counter Kelvin who said, rather sarcastically, that they would have to discover a new, hitherto unknown source of heat if they wanted an Earth that was older than the one he was supplying them with.

Then, in 1896, radioactivity was discovered. Uranium atoms were slowly breaking down and giving off energetic radiations. So were thorium atoms and a few other varieties of massive atoms. In 1901 it was definitely established that radioactive atoms, in the course of their breakdowns, were giving off heat.

A young New Zealander, Ernest Rutherford, who was devoting himself to work on radioactivity, considered the matter. Each radioactive atom in the Earth's crust liberated only an infinitesimal amount of heat as it broke down. If, however, *all* those that broke down were taken into account, the total amount of heat liberated was enormous. It would be enough to keep the Earth cooling down very slowly indeed.

In other words, the Earth could have been in existence for billions of years without cooling down any more than it has. And as for the Sun, perhaps it *didn't* shrink. Perhaps it, too, made use of the heat derived from radioactive phenomena.

In 1904, when Rutherford was thirty-three, he addressed a scientific body on the subject—and there in the audience was the eighty-year-old Kelvin. Rutherford did not enjoy the thought of facing the giant and being forced to contradict him and he hoped the old man would fall asleep. Kelvin did not. As Rutherford approached the crucial point of the argument, Kelvin's eyes fixed themselves balefully on him.

Carefully, Rutherford pointed out that in advancing his own arguments, Kelvin had said his conclusions were correct unless some hitherto unknown source of heat were discovered. Well, said Rutherford, Kelvin's amazing foresight had been vindicated. A new source of heat *had* been discovered, and here it was. Thereupon, Kelvin's face relaxed into a smile and Rutherford was home safe.

And so were the geologists, who found they had an Earth billions of years old to play with.

# 43

# Scientists Are Human

An old-fashioned stereotype of scientists is that they are cold, reasoning machines, immune to emotion. How wrong that is! Scientists are human and, however ingenious their brains, however powerful their thinking processes, they can be hurt as often as anyone else. Their emotions can rule them, and they can know misery and despair.

The French mathematician André Marie Ampère (1775-1836) was eighteen when his beloved father, a well-to-do merchant, was guillotined during the French Revolution. Young Ampère fell into a profound depression as a result. But then he fell in love and married a young woman he adored—but she died in 1804, after they had been married only a few years. He never recovered from that. Still, Ampère made enormously important discoveries in electricity. He founded the discipline of electrodynamics (the quantity of electric current is measured in "amperes" in his honor), but his scientific success did not make up for his personal tragedies in his mind. When he died, the epitaph on his gravestone was one he had chosen in advance: *Tandem felix,* which means "Happy at last."

Ludwig Edward Boltzmann (1844-1906), an Austrian physicist who, with James Clerk Maxwell, had worked out the kinetic theory of gases, committed suicide after episodes of severe mental depression. There is the possibility that part of his misery arose from brooding over the savage criticism of his (perfectly correct) kinetic theory by other scientists. (Isaac Newton couldn't bear criticism either. He didn't kill himself over it, but he did have a serious nervous breakdown.)

Emil Hermann Fischer (1852-1919) was a German chemist who worked out the structure of the various sugars and established their stereoisomerism. He then went on to work out the chemistry of the purines, and in 1902 he won the Nobel Prize for chemistry. He was an ardent German nationalist, and during World War I he organized German food and chemical production for the war. Two of his three sons died in the war, which ended in German defeat. Then, when the sorrowing Fischer found he had cancer, he did not wait for death, but killed himself.

Hans Fischer (1881-1945) was no relative of Emil, but in his younger days he served the latter as assistant. The lives of the two Fischers were curiously similar. Hans Fischer worked out the structure of porphyrins, an important grouping in hemoglobin and chlorophyll, and he won the Nobel Prize in chemistry in 1930. Then came World War II. Again Germany was defeated, this time far more catastrophically. Shortly before the end of the war, air raids over Munich destroyed Hans Fischer's laboratories. In despair, he killed himself.

Other suicides include such American scientists as Bertram Borden Boltwood (1870-1927), who first showed how to measure the true age of the Earth; George Eastman (1854-1932), who pioneered photography for the masses; and Percy William Bridgman (1882-1961), who worked with high pressure and was the first to produce artificial diamonds.

Whatever a scientist may do and be, he is in all things human as subject to woe and misery as any artist or bookkeeper or housewife.

# 44

# Sometimes It Takes Time

Sometimes it takes time to get the honor one deserves.

Back in 1909, an American physician named Francis Peyton Rous joined the staff of the Rockefeller Institute for Medical Research. He was thirty years old at the time.

Not long after he began work at the Institute, a poultry breeder wandered in with a sick Plymouth Rock chicken that he wanted examined. It had a tumor, and when it died, Rous decided to test whether it might contain a virus. (He was sure it didn't.)

He mashed up the tumor and passed it through a filter that would keep out any infectious agent but a virus. He found, however, that this "cell-free filtrate" was infectious and would produce tumors in other chickens. He did not dare call it a virus in the report he published in 1911, but as time went on and more and more came to be known about viruses, it seemed there was nothing else to call it. In the 1930s, it became known as "the Rous chicken sarcoma virus," and it was the first of the "tumor viruses."

More time passed. Viruses were found to be nucleic acids coated with protein. The nucleic acid part could penetrate cells and could sometimes add on to the chromosomes, which were also nucleic acids coated with protein. In this way, viruses could alter cell chemistry and potentially induce tumor growth.

By 1966, the importance of such virus activity was clearly seen to make Rous's report of fifty-five years before worth a Nobel Prize. You can't give Nobel Prizes to dead people, of course, but by great good fortune, Rous was still alive. He had just passed his eighty-seventh birthday, but he was still actively at work in his laboratory. He accepted his Nobel Prize, having waited over half a century for it. He was the oldest man ever to receive one. He died in 1970, four months past his ninetieth birthday.

Consider Henry Gwyn-Jeffreys Moseley, on the other hand. He was an English physicist who, in 1914, at the age of twenty-seven, studied the X rays emitted by metals under certain circumstances and evolved the notion of "atomic number."

This, for the first time, made real sense out of the periodic table, pointing the way toward determining the structure of the atomic nucleus.

The importance of the work was recognized at once and Nobel Prizes were handed out to a number of scientists who worked on the concept. For instance, a Swedish physicist, Karl Manne Georg Siegbahn, carried on Moseley's work, showing how to handle the X rays more precisely. He got a Nobel Prize in physics in 1924.

Moseley, however, did *not* get a Nobel Prize.

The trouble was that World War I started in 1914 and Moseley enlisted at once as a lieutenant in the Royal Engineers. In later wars, Moseley's brain would have been recognized as too valuable to risk, and he would have been placed in a laboratory to work on war research. In World War I, however, military leaders were stupid enough to send him to the front.

On August 10, 1915, Moseley was killed in the course of the badly mishandled Gallipoli campaign, while he was still only twenty-seven.

It seems quite likely that he would have gotten the Nobel Prize before he was thirty, but in his case that was too long to wait for the honor he deserved. He didn't make it.

# 45

# Learning Science

I imagine that many a young scholar has asked him- or herself, rebelliously, why on Earth s/he must learn science when s/he has no intention of being a scientist.

Someone who feels that may feel that s/he need know no more than the minimum that will allow him or her to just barely get through life. Why should one know history if one is not going to be a historian? Or geography or languages if one isn't going to travel much?

But surely there is more to life than what one "does." Even if one lives quietly at home and works at some simple, routine job, there must nevertheless be *some* value to understanding the world about us, to understanding events in the light of the past, to having an appreciation of other places and other cultures.

In fact, it is surely *fun* to know things. It brightens one's life, sharpens one's wits, reduces one's boredom, broadens one's horizon, makes one more interesting and more pleasurable to be with.

This is true of any sort of knowledge or skill, actually, even of those that are not strictly "school subjects." Someone who knows how to carve wood into clever little devices, or who knows all about stamp collecting, is surely more fun to be with and to watch and to listen to than someone who knows nothing at all.

If, then, you know these other things, do you have to know science, *too?* Is there something special about science?

Actually, there is.

Our modern world is founded on science—and on technology, which is the application of science to everyday affairs. Almost everything we do depends on our modern devices, such as automobiles, record players, and television sets, and these in turn depend on scientific principles. Our future will depend on computers, robots, nuclear power, rocket ships, all of which only make sense if we understand science.

If a person does not understand what makes these things work, they might as well be magic. People without science live in a mystery world that makes no sense to them. Even if they say, "So what? All I want to do is make a living, have a family, and look at the scenery," they may find that is not so easy. In an increasingly scientific world, the good jobs, the money-making jobs, will go to those who understand science.

Then, too, science has its dangers and its benefits. Used improperly, science can flood the Earth with pollution, with dangerous chemicals, with radiation, with devices that destroy our privacy and our freedom. Used wisely, however, science can increase our energy and food supply, improve our health, expand our joy, extend our lives, and broaden our sense of security.

Who decides how best to use science, however? In a democracy, it should be the people generally. But how can the people come to an intelligent decision if hardly any of them know much about science to begin with?

Surely it will be increasingly important, as the years pass, for people to understand science if they are going to be expected to help make intelligent decisions about how to use science to save the world, and not destroy it.

That is why it is important to study science, even if one is not going to be a professional scientist.

# 46

# Self-Correcting

Every once in a while (not often), scientists discover that one of their number has published false data or has plagiarized someone else's work.

This is always deeply embarrassing, especially since in these days the news usually receives wide publicity in the nonscientific world.

In some ways, however, these scandals actually reflect credit upon the world of science. Consider:

1. Scientists are, after all, human. There is enormous pressure and competition in the world of science. Promotion and status depend on how much you publish and how *soon* you publish, for the lion's share of credit comes if you are *first* with an important theory or observation. Under those circumstances, there is a lot of temptation to rush things, to make up some supporting data you are sure you will eventually find anyway, or to help yourself to someone else's work. The surprise, really, is not that it sometimes happens, but that it doesn't happen much more often. Scientists, almost unanimously, resist the pressure marvelously well.

2. When it does happen, the mere fact that it is so publicized is a tribute to scientists. If it were a common event, or if people expected scientists to be corrupt, it would make smaller headlines and drop out of sight sooner. Single cases of scientific corruption, however, will be talked about for years and inspire articles and books by the score.

3. Cases of scientific misbehavior point up the actual difficulty of carrying them through successfully, or even for very long. In fact, most cases of such misbehavior occur in the biological and medical sciences, where data and theories are less elegant than in the physical sciences. Animal behavior and tissue chemistry are simply less neatly organized than the movements of stars and atoms or the flow of energy, and something that isn't so in the former case is less easily detected. Yet a vital principle in scientific research is that nothing counts until observations can be repeated independently, and there, almost inevitably, anything peculiar is uncovered. Science is *self-correcting* in a way that no other field of intellectual endeavor can match.

4. It is scientists themselves who catch the frauds; no one else is equipped to do so. The point is that scientists *do* catch them. There is never any cover-up on the grounds that science itself must not be disgraced. However embarrassing the facts may be, the culprit is exposed pitilessly and publicly. Science is *self-policing,* and *effectively* so, in a way that no other field of intellectual endeavor can match.

5. Finally, the punishment is absolute. Anyone who proves to have broken the ethics of scientific endeavor is ruined for life. There is no second chance, no remains of status. He or she must drop out, forever disgraced.

Add to all this the fact that scientific ethics requires all scientists to labor to find flaws in their *own* observations and theories, and to publicize these flaws when they find them, and you will understand how stern science's requirements are. Perhaps it is not so astonishing that scandal comes so infrequently.

# 47

# The Knowledge of Good and Evil

There's no denying that the findings of science can prove dangerous. Is it safe to know about nerve gases? Aren't we better off to remain ignorant of sophisticated space weapons? Is there deadly danger in recombinant DNA? Ought we to have learned to split the uranium atom?

Well, then, ought we to set up some agency to guide us, to direct and limit scientific research, to say, "Thus far and no farther"?

I think not, for two reasons.

First, we must distinguish between knowledge and misuse. To know that DDT affects insects in certain ways can allow us to better understand insect biochemistry, and then perhaps our own. Worthy uses may then stem in many directions.

To use DDT indiscriminantly and without adequate testing or forethought could, on the other hand, do vast ecological damage.

To understand the fact of uranium fission could help us penetrate the workings of the Universe; to use uranium as an explosive and to set it off in anger may end by destroying civilization.

This is not a new phenomenon. We have always faced it. To understand how to start a fire, to study its effect on food, clay, sand, and ores yields fascinating knowledge without which civilization could not have been erected.

To use fire to burn forests, structures, and heretics, or to use it under any circumstances without adequate ventilation, can create untold misery.

Let us by all means guide and direct the *use* of knowledge, but not the gathering of knowledge itself.

But is the gathering safe? If human beings know something that can be put to evil use, will not the temptation to do so, sooner or later, be overpowering? And would it not be safe simply to remain in ignorance? Except that *anything* can be put to evil use, and that ignorance can also be dangerous.

My second reason is that it isn't possible to distinguish, surely, between good and evil.

Who objects to medical advances? The discovery of anesthesia, vitamins, hormone therapy, and new techniques of surgery is greeted with universal applause. And of all the medical discoveries, the greatest was Louis Pasteur's development of the germ theory of disease in the 1860s. The germ theory led to the rapid control of infectious disease. It limited, inhibited, aborted, and all but ended the deadly plagues and epidemics that threatened humanity through all its history. It contributed mightily to the doubling of the average life-span from thirty-five to seventy in the last century and a quarter in those parts of the world where modern medicine exists.

But nothing has fueled the population explosion like that rapid decline in the death-rate. And it is overpopulation that now threatens the world more than anything else does. How the teeming billions contribute to the consumption of resources, the production of pollution, the destruction of the land, and the building of the frictions and tensions that produce alienation, violence, and ultimately, perhaps, nuclear war.

Is modern medicine good or evil then? It has saved millions of lives, but might it not end by destroying billions? Would you have stopped Pasteur when he worked out the germ theory? Or not?

# 48

# Science and Technology

It's so easy to make the distinction. Science is "basic," "pure," "intellectual," "good." Technology is "applied," "derived," "commercial," "bad."

It all started with the Greeks, where philosophic speculation was the prerogative of free men who didn't have to grub for a living because they had slaves to do the work for them. Part of the philosophic speculation was on the nature of the Universe, the movement of bodies, the properties of matter and of life—all that we call science today.

As soon as knowledge was applied to daily life, however, so that something had to be carved or hewn or hammered or wedged or put together—it became *work,* fit only for artisans and slaves.

Greek science reached a dead end, for if the only thing that counts as fitting is to fold one's arms and speculate, one ends up with nothing, really. Democritus speculated there were atoms and Aristotle speculated there were not, and everyone was free to choose between the prettiness of the speculations and decide what they wanted to believe—and Aristotle won.

By the sixteenth century, though, it was decided that the Universe will answer questions only if you put in a little work. You have to earn the answer by a little physical sweating. You have to set up experiments.

With the coming of experimental science, scientists had to play about with balls and inclined planes, and they had to devise new instruments with which to make measurements. The lofty speculation of free men and the handiwork of artisans and slaves became mixed up. Inextricably.

In the 1590s, Galileo studied a swinging chandelier in a cathedral and came up with the principle of the pendulum. He checked it out by a little artisanry, however. When he got home, he set up pendulums of different lengths swinging through differently sized arcs and convinced himself.

It would have been easier if Galileo had a good timepiece to measure the time of swing, but no such thing existed until the 1650s, when Christian Huygens invented the first modern clock—using what? Why, using the principle of the pendulum.

And once the clock existed, what did that make possible? Why, it revolutionized the study of astronomy, by making it much easier to time the motions of the heavenly bodies. Thus, it helped the purest of sciences to advance.

In fact, science would stop dead were it not for the continuous drizzle of ingenious inventions that make its advance possible. Our new knowledge of the planets was made possible by what? By rocket-powered probes, built by technology.

Even the purest of the pure, mathematics, is contaminated. The four-color problem of mapping—which has, as far as we know, no practical uses whatever—was solved by what? Why, by computers, built by technology.

Yet we still talk about science and technology as though they were differnt things, although for four centuries they have lived together in close symbiosis, each unable to exist without the other.

It's time we stopped.

# 49

# Missed Opportunities

I suppose that, as the grand march of science proceeds, every science fiction writer has a chance to bemoan missed opportunities.

Forty years ago, I thought of writing a story about a very tiny but massive star around which a planet circled in a two-minute period of revolution. In other words, I had thought of a neutron star, decades before any had been discovered—and then I didn't write the story.

About the same time, I actually started a story about a star so large and dense that light could not escape from it—a black hole, though the expression was not to exist for decades—and I never finished it.

I've been regretting those incidents for many years, and now I have a new cause for regret.

The *Voyager* satellites have uncovered fascinating details about Saturn's rings, details that couldn't possibly have been guessed from observations here on Earth, about one and a half billion kilometers away. There are the braided ring, the so-called "spokes," the manner in which small satellites "herd" some of the rings and keep them from dissipating. We couldn't have guessed these matters, since even now that we know them, we still can't explain them.

The matter of the sub-rings, though, we might have guessed. We can see the largest of all the ring separations, the "Cassini division," from Earth, but some astronomers have claimed to see other finer divisions, the existence of which always seemed dubious. But why not *expect* other divisions? Why not take it for granted that, from a close view, the rings would show much more detail than they would from one and a half billion kilometers away?

I had a rare chance to make the guess myself. Back in 1952—over three decades ago—I published a story called "The Martian Way." In the course of the story, a group of Martian colonists traveled to Saturn. That gave me the opportunity to describe the rings as seen at close quarters, and I did it this way:

> *[The rings] emerged from behind Saturn, a tight, bright triple band of orange light. . . . They widened as they came, like the flare of a horn, growing hazier as they approached, until, while the eye followed them, they seemed to fill the sky and lose themselves.*

> *. . . the rings broke up and assumed their true identity as a phenomenal cluster of solid fragments, rather than the tight solid band of light they seemed.*
>
> *Below him . . . was one of the ring fragments. . . . Other fragments were farther off, sparkling like star-dust, dimmer and thicker, until, as he followed them down, they became rings once more.*

It was all accurate enough, but I clung slavishly to what astronomers saw from a vast distance, never once grasping that my astronauts, from their position within the rings themselves, were bound to see more detail and finer divisions.

Suppose I had said, concerning the rings:

> *. . . They emerged from behind Saturn, a tight, bright triple band of orange light. . . . They widened as they came, and, as they approached, finer divisions became visible, so that the rings seemed made up of narrow, parallel, curving stripes of light, separated by darker hair-lines. . . .*

Oh, if I had only said that. It would have been a reasonable guess, and I would now be basking in the glory of having been the person who had gotten it *right.*

But what's the use? I wasn't smart enough to have seen it, except in hindsight.

# Part V

## "Foreword by Isaac Asimov"

I think that I have written more book introductions and forewords than anyone else in the world. For one thing, I have written introductions to nearly all the various anthologies I have edited or coedited, and there have been over a hundred of those. For another, I frequently write introductions to books in whose production I have not at all been involved—provided, of course,I see the book in manuscript and enjoy reading it.

Actually, writing introductions is pleasurable, for my idea of writing one is not merely to praise the book. Rather, I seek in the theme of the book some related angle that, in my opinion, is not elsewhere adequately discussed, or not discussed at all. That gives me a chance to talk about matters I might not otherwise have thought to discuss, and that invariably brightens my day.

Naturally, I usually end with *some* mention of the book itself, and I praise it, for I would not write the introduction if I did not think the book deserved praise. Such praise might seem out of place in a collection like this, but I have sometimes hesitated to cut it out. It's nice to say good things about a good book.

In any case, here are forty-two introductions that I have written. It is only a quarter, perhaps, of all those I have written, but it is quite enough to give you a taste of the variety, I think. (As a matter of fact, I may have one or two introductions elsewhere in other sections, but I warned you that my division into categories was a rough one.)

# 50

# Shuttle

The word "shuttle" goes back to the old Norse word *skuttil,* meaning "arrow" or "harpoon," a missile that travels a relatively short distance—from the person who hurls it to the target it strikes. The word "shoot" comes from the same source, and my own feeling is that the ultimate origin is onomatopoieic—by sound.

A narrow missile making its way rapidly through the air makes a sound like "shoosh," "woosh," or "s-s-s-t," depending on the subjective interpretation of the person doing the hearing. From that, through a variety of changes, comes "shuttle" and "shoot."

In prehistoric times, "shuttle" developed a markedly peaceful meaning—a development that always seems pleasant to me.

The art of weaving was invented before the days of writing, so we don't know the origins. Like most prehistoric inventions, later peoples, unable to imagine how mere human beings could think of such a clever thing, assumed that some god had granted the knowledge.

In any case, to weave you must begin with a series of parallel vertical threads, strong and tightly woven. (This is the "warp.") However, those parallel threads must be held together, and to do that, you place another series of threads, somewhat thinner, at right angles (the "woof"). Each thread of the woof goes over one thread of the warp, under the next, over the next, and so on. The next thread of the woof goes under, over, and under, so that no two adjacent threads of the woof both go under or both go over a particular thread of the warp.

If the threads of the warp and the woof are then shoved close together, you have a two-dimensional set of threads that hangs together and is much stronger on the whole than the individual threads are. You end with a piece of cloth, a textile material—light, porous, and much more suitable for keeping the body protected, clean, and comfortable, at ordinary temperatures, than furs or leather are.

The catch is that threading the woof along the warp takes a long, long time. Therefore (still in prehistoric times), the "loom" was invented. The threads of the warp were attached to each of two portions of the loom, which could be pulled apart. This meant the threads of the warp could be alternately in, out, in, out. The threads of the woof could then be pushed through in a straight line, and when the pieces of the loom were joined together again, the woof would cross over, under, over, under the warp.

To increase the speed of operation, the woof is tied to a heavy object that is hurled across the width of the warp. This object is the "shuttle." After all, it is a missile and it travels a short distance. The shuttle carries the woof across the warp and then (with the loom separated in the opposite sense) back across the warp, and then forward again, and then back, and so on, the loom always being manipulated so that the woof goes over and under the warp as it is supposed to.

Because of the backward and forward motion of the shuttle during weaving, the word "shuttle" gained a new refinement of meaning. Not only was it a missile that traveled a comparatively short distance, but it was one that went ceaselessly back and forth over that short distance.

Thus, if some means of transportation travels from point A to the not very distant point B, then back to point A, then back to point B, and so on, indefinitely, *it* is a shuttle, too.

In Manhattan, for instance, there is a "subway-shuttle" that travels endlessly from Times Square to Grand Central, a distance of just about two-thirds of a mile, which is short considering that a number of subway lines in the city are twenty miles long or so. Naturally, people tend to economize on syllables, and the subway shuttle becomes simply "the shuttle" to subway users.

Similarly, once air travel became common, there were bound to be numerous flights between big cities that were quite close together. Thus, there would be a dozen flights or so each day between New York and Boston, or between New York and Washington. These are trips of about two-hundred miles, as compared with some flights that are anywhere from three thousand to ten thousand miles long. One speaks, therefore, of an "air-shuttle;" those who often travel by air speak of "the shuttle."

The next extension is obvious to science fiction readers. Our first attempts in space were to place objects in orbit about the Earth or to send an object to the Moon and back. After that, we began to think of building a space station in orbit about the Earth. It could be permanently occupied (in shifts) and serve as a base from which we could carry on long-term experiments, build space structures, and launch spaceships on long journeys.

Naturally, we would need some kind of space vessel that would specialize in traveling from Earth's surface to the space station and back, in order to bring in supplies, take off wastes, replace personnel, and so on. Since the space station would be considerably closer to the Earth than the Moon is, such a space vessel would be traveling a comparatively short distance—up, down, up, down, indefinitely. What would it be called? Inevitably, it would be a "space-shuttle" or, to space-hounds, "the shuttle."

Actually, the United States built a vessel suitable for such use long before it built a space station that could profit by that use. The first space shuttle, named *Columbia,* was launched on April 12, 1981, just twenty years to the day after the first human being, Yuri Gagarin, had orbited the Earth. Four shuttles were eventually put in operation, and by January 1986 some two dozen flights had been made, with triumphant success.

Science fiction writers can write stories about such shuttles or about any devices that could represent comparatively short flights endlessly repeated. If we ever learn the trick of hyperspatial travel, we might conceivably talk of "the shuttle" between the Sun and Alpha Centauri, or between the Milky Way and Andromeda. We might also have a shuttle between one end of a cylindrical space settlement and the other. The changes that can be rung, as you can see in the stories we have collected in *Space Shuttle,* are endless.

Of course, on January 28, 1986, the space-shuttle went sour (only temporarily, I hope) with the explosion of the *Challenger.* But, you know, disasters of this sort, however horrifying and deplorable, are inevitable in any great and risky project.

My story "The Last Shuttle," included in this book, was written in March 1981, a month after the first space-shuttle flight, and was published in a Florida newspaper soon after being written. In it, I have a character point out that space-shuttle lift-offs are not necessarily trouble-free. "There was the case of Enterprise Sixty," she says, referring to an episode early in space-shuttle history.

Well, I got the name wrong, but I saw it coming, five years in advance. . . .

# 51

# The Good Deed of *Voyager 2*

*Voyager 2* is the most successful unmanned planetary probe that human beings have ever sent out. This is not to knock any of the others: those that landed on Venus and scanned Mercury and mapped both planets (the first by radar rather than light); those that mapped Mars in detail, to say nothing of landing on Mars and testing its soil chemically; and those that preceded *Voyager 2* in passing through the asteroid belt and beyond in order to sail into the vast void in which the outer planets circle.

None, however, did quite as spectacular a job as *Voyager 2.* That probe, in a flight of nine years, managed to skim by both Jupiter and Saturn, and then move on to Uranus, where it made observations of a giant world so far from Earth that the ancients didn't even know it existed. It gleams in the sky just barely brightly enough to see as a very dim star.

And *Voyager 2* is not yet done. For several additional years, it will continue to forge through space until it passes Neptune, the farthest major planet we know. (Pluto is farther most of the time, but it is an icy pipsqueak, smaller than our Moon.)

This long flight has been full of adventure, of narrow escapes, of seeming disasters somehow averted or righted. I won't try to tell the story here. Joel Davis tells it in full and in all its drama and excitement in *Flyby.*

What I intend to talk about, however, is the good deed of *Voyager 2,* something that it has done that no one could possibly have foretold or foreseen.

Let us backtrack. Human history is full of disasters, many of them beyond any possibility of human control. There are endless tales of floods and storms, earthquakes and volcanic eruptions, droughts and pestilence, every one of them bringing endless human suffering. There are human-sponsored disasters, too, such as wars. Those of us who are past middle age can remember World War II, the most disastrous war ever fought, and we all look forward with dread to the possibility of another war that might easily end the human race.

Even those human achievements we can be proudest of take their disastrous toll. No great bridge has been built, no great skyscraper, no great tunnel, without its deaths. The automobile, which no one would dream of giving up, kills fifty thousand a year in the United States alone. We are inured to air disasters, to deaths in mines, in factories, from chemical poisons, from radioactive leakage.

It follows, then, that few of us can really expect the exploration of space to be disaster-free. We *know* it will cost lives, and it has. The United States lost three astronauts on the ground in a space-capsule fire in 1967. The Soviet Union lost three astronauts when a leak developed even as they were getting ready to return to the atmosphere.

What's different about space exploration, however, is that the whole world is watching. In December 1985, a plane crashed in Newfoundland and killed over 250 soldiers. No one watched the crash. The report came after the fact, and that made it merely a news item—like others. The nation took the blow, shook its collective head, and, except for the immediate families of those lost, went about its business. Disasters do take place.

A little over a month later, however, on January 28, 1986, the shuttle *Challenger* was launched, with a crew of seven. This was being watched by millions. And there, in front of all those eyes, a minute and half after launch, it turned into a fireball. All those millions realized they had just watched seven brave people have their lives snuffed out. To make it worse, one of the seven was a teacher, the first "ordinary" person to go into space.

Nothing has ever shaken the United States, and, indeed, the world, as has that *witnessed* disaster. Not even the assassination of an American president struck so many into a state of shock that lifted with such difficulty.

We all know what that has done to the American space program: the self-flagellation of "what did we do wrong?"; the determination to make everything ten times as safe before we move again; the postponement and further postponement of everything.

It may seem difficult to imagine how anything could have made the situation worse, but it can be done. Suppose that *Voyager 2* had not just passed Uranus at the time. Suppose that the results, spectacular and exciting, weren't coming in.

By the sheerest chance, we had one of the great *successes* of the space program playing before our eyes, just as the *Challenger* disaster struck. Don't get me wrong. Nothing that *Voyager 2* could have done, or discovered, could possibly have given us back those seven lives. It couldn't have given us back our lost confidence, or made up for the shock and horror of it all.

It did show us, however, at just the time we needed it most, that space exploration could provide success as well as disaster. *That* was the good deed of *Voyager 2* that no one had foreseen. Without the example of *Voyager 2,* it is possible that the *Challenger* disaster might have plunged us so incredibly deep in disillusion and depression that our American space effort would have been foreclosed forever, or at least for so long that we would have left space to others—to the Soviet Union, to western Europe, to Japan.

As it is, *Voyager 2* provided just the unexpected push that may make it possible for us to recover in time.

We must realize that all progress toward great goals requires us to evade or overcome the obstacles that inevitably stand in the way. The highways to success are not smooth and never have been smooth, and they are paved with heartbreak.

*But we must not give up.* Disasters like that of *Challenger* will come again, but successes like that of *Voyager 2* will also come again.

# 52

# The Longest Voyage

Suppose you want to take a trip across the country, from Portland, Maine, to Portland, Oregon. That's roughly 3,000 miles. A trip around the world along the Equator is only a little over eight times that, 25,000 miles.

To go from the Earth to the Moon is only about nine times the Equatorial jaunt, about 240,000 miles. Beyond that? Well, Venus at its closest is just over a hundred times the distance of the Moon; it is about 25,000,000 miles away. And right now, Pluto is just about as near to Earth as it ever gets, but it is over a hundred times farther away than Venus is—more than 2,800,000,000 miles away.

So far we've stayed in our Solar System, but beyond that are the stars. Even the nearest star is nearly 9,000 times as far away as Pluto is right now. The nearest star is Alpha Centauri, and that is 25,000,000,000,000 miles away. And that's the *nearest* star.

The distance across the Milky Way Galaxy is 23,000 times the distance from Earth to Alpha Centauri. The distance from here to the Andromeda Galaxy, the nearest large galaxy to our own, is about twenty-three times the diameter of the Milky Way Galaxy. And the distance from here to the farthest quasar is about 4,000 times that.

What about time? It takes a few days to get to the Moon, a few months to get to Venus or Mars, a few years to get to the giant planets of the Solar System. But that's about as far as we can go and have it make reasonable sense.

To get to even the nearest star, at the present state of the art, would take hundreds of thousands of years. All that NASA has done so far in sending probes as far as Saturn has been to play games in our own backyard. It is *interstellar travel,* trips to the stars, that represents the longest voyage.

And it is in trips to the stars that science fiction writers and readers are most interested. Our Solar System is too well-known and too limited. The Solar System (outside Earth) is not at all likely to bear life of any kind—certainly not intelligent life. So we've got to take the longest voyage and get to the stars, if we're to find extraterrestrial friends, competitors, and enemies. As long ago as 1928, in *The Skylark of Space,* E. E. (Doc) Smith took the first science fictional trip to the stars—and how the readers loved it!

Good old Doc was a little vague on just how his interstellar ships managed to cross those huge spaces, however, and, to tell you the truth, we're not much better off now. Let's list the possibilities:

1. We can keep accelerating, going faster and faster and faster until we're going fast enough to cover vast interstellar and intergalactic distances in a matter of months, or even days.

Objection: Physicists are strongly of the opinion that the speed of light in a vacuum, 186,262 miles per second, is as fast as anyone can go. At that speed, it will still take years to reach the nearest star, millions of years to reach the nearest large galaxy.

2. Even if we're limited to the speed of light, that could be good enough. As one approaches the speed of light, the rate of time-passage on the speeding

object slows steadily, and at the speed of light itself the rate of time-passage is zero. At light-speed, then, the crew of a starship would cover enormous distances practically instantaneously.

Objection: Interstellar and intergalactic space is littered with occasional hydrogen atoms. At light-speed, these atoms would strike the ship with the energy and force of cosmic ray particles and would quickly kill the starship's crew and passengers. The ship reasonably would have to go no faster than perhaps one-fifth light-speed, and at that speed the time effects are not great enough to help us much.

3. Suppose we attach a kind of "atom-plow" arrangement in front of the starship. It would scoop up all the atoms in front of it, thus preventing cosmic ray problems and, in addition, gathering material to serve as fuel for its nuclear-fusion engines.

Objection: Such atom-plows would have to be many thousands of miles across to be effective. Building such things would represent enormous and perhaps insuperable problems.

4. We can evade the speed-of-light limit altogether by making use of tachyons, subatomic particles that move much faster than the speed of light and that, as a matter of fact, cannot move slower than the speed of light.

Objection: Tachyons exist only in theory; they have not actually been detected. Most physicists think they will never be detected. Even if they were detected, no one has even come close to figuring out a way of putting them to use.

5. Perhaps we can evade the speed-of-light limit by going through black holes. They, at least, are known to exist.

Objection: Even if black holes exist (and astronomers are not yet unanimous on this), no one is even close to suggesting how any starship might approach one without being destroyed by tidal forces. In addition, there is by no means general agreement that one can negotiate long distances quickly by going through black holes.

6. In that case, we might find some other way of leaving this universe. We could travel through hyperspace in "jumps" that will carry us enormous distances in zero time.

Objection: So far, hyperspace exists only within the imagination of science fiction writers.

7. Well, then, we can submit to the speed-of-light limit, but freeze the crew and passengers, and arrange to have them restored to conscious life after thousands of years have passed and the destination has been reached.

Objection: No one really knows how human bodies can be frozen without being killed, or whether such frozen bodies, even if they can retain a spark of life, can retain it over thousands of years.

8. In that case, there seems nothing left to do but to coast—to travel at ordinary speeds, considerably less than that of light, with all people

aboard thoroughly conscious. This means it will take many thousands of years to reach even the nearer stars, so that many generations will have to spend their lifetimes aboard the starship. That may be bearable if the starship is large enough.

Objection: None, really, if people want to do it.

* * *

So much for hard-headed realism. In science fiction, we tend to have faith that problems that seem insuperable now will be solved—perhaps in ways that are utterly unexpected.

Therefore we are offering you a baker's dozen of stories, all involving starships. In these are explored the various strategies I have described above for covering long distances, and perhaps one or two that are too far out for me to have even mentioned.

What's more, the stories explore the effect of the long voyages on the people on board the starship, and the kind of events that might take place on them.

Since it is not likely that such voyages will be undertaken in our lifetime (and certainly not completed, if the generations-long coasting starship should indeed prove to be the only practical alternative), these exciting science-fictional speculations are the only way we can experience, if only vicariously, the long voyages that are the quintessential dreams of the far-flung imagination.

# 53

# Spreading Through Space

How did life originate? Biologists have not yet managed to work out a definite scenario. Indeed, the task seems so formidable that there would be a kind of relief in sitting back and saying, "It never happened."

And yet it *did* happen. Here we are. Presumably, very simple compounds existing in the ocean and atmosphere of the primordial Earth, making use of ultraviolet light from the Sun, lightning, volcanic heat, or other sources of energy, gradually built up into more complicated compounds and eventually developed properties we recognize as those of simple life.

The question is whether conditions on the primordial Earth were such as to allow this sort of thing to happen. Even if such conditions did exist, it is possible that the requirements are so tight that only on very few worlds could life develop.

But perhaps very few, or even one world is sufficient. Although it may not be possible for a world to give rise to life, such a world may be hospitable to life that formed somewhere else and arrives there.

Thus, although life couldn't possibly have developed on dry land, having developed in the sea and lived there for three billion years, such life was capable of colonizing the land. Again, though life couldn't possibly have developed on Antarctica or in the Gobi Desert, having started elsewhere, it could spread into those inhospitable areas.

Is it possible, then, that life may start on some particularly suitable planet and then spread through space to colonize other planets? Might we be a starting point, and might there be other planets, less favored than our own, that have life distantly related to ours? Or might life have started elsewhere on a world *more* favorable than Earth, and might it eventually have colonized our planet?

The first scientist to take up the notion of life from one planet colonizing another was a Swedish chemist named Svante August Arrhenius (1859-1927). In 1907, he published *Worlds in the Making,* picturing spores escaping by random movement from the atmosphere of a planet and then being driven through space by the light-pressure of the planet's sun.

Driven this way and that by starlight, eventually the spores might travel randomly for many millions of years until by sheer chance they would strike the atmosphere of a suitable planet and seed it.

At first blush, this theory looks attractive. Bacterial spores, protected by a thick coat, are very resistant to cold and dehydration and might conceivably last a long time in the vacuum of space. Moreover they are just the right size to be more affected by the outward pressure of a star's radiation than by the inward pull of its gravity. To be sure, in 1910 it was found that the bacterial spores we know of are vulnerable to ultraviolet light—not to speak of other destructive radiations, like cosmic rays, stellar X rays, and magnetospheres full of charged particles.

Perhaps, though, there are some spores resistant to radiation, or perhaps microscopic forms of life arrive at a planet by some means other than simple chance. The English scientist Francis H. C. Crick (b. 1916) has suggested—possibly half in fun—that an exploring party might land on a planet that is lifeless but is capable of supporting life, if life is introduced. They may leave their garbage behind, the remains of their lunch, let us say, and the microorganisms present may survive and start the vast cycle of life. In other words, the simple cells may have made the voyage under protected conditions.

As a matter of fact, since 1968 astronomers have detected molecules in interstellar clouds—including some surprisingly elaborate ones, too—mostly carbon-containing (organic) molecules. One molecule containing as many as thirteen atoms has been located. The more elaborate the molecule, the fewer there are of them and the more difficult they are to detect. It may be that if we can only get close enough and study the clouds delicately enough, molecules on the direct route to life might exist there. In that case, we might then hypothesize that a planet can be seeded by these clouds.

To be sure, it is hard to see how material from an interstellar cloud can get to planets. We do know, however, that comets, which are part of planetary systems, can be partially composed of carbon-containing molecules, and some of these may be complex, too. Indeed, in 1986 a close study of Comet Halley showed that it contained long polymeric molecules.

Not only that, but there are certain rare meteorites called "carbonaceous" meteorites, which contain small quantities of water and organic molecules. Some of these meteorites have been analyzed and have been found to contain fatty acids and certain amino acids that are the building blocks of proteins. The nature of the compounds present show that they were not formed through the agency of life, but they might still serve to seed a planet and get bigger things going.

Perhaps the most radical ideas held in this connection are those of the English astronomer Fred Hoyle (b. 1915) and an Indian colleague, Chandra Wickramasinghe. They feel that the syntheses in interstellar clouds and in comets may go far beyond what has been detected. Very small quantities of microscopic bits of life might have been formed, they maintain. They suggest that life on Earth might have originated when spores were carried to Earth by comet tails—and in recent times, new and virulent pathogens (like the virus of the 1918 influenza pandemic) may have been deposited on Earth in this manner. (It is only fair to say, however, that the Hoyle-Wickramasinghe notions are not taken seriously by other scientists.)

Nevertheless, the concept of extraterrestrial seeding, while an excellent plot-device for science fiction, is not science fiction only. As you will see in John Barnes's *Sin of Origin,* serious scientists have canvassed the possibility.

# 54

# First Contact

In science fiction parlance, "first contact" refers to an initial meeting between Earthpeople and some extraterrestrial intelligence. It is something that has never yet happened, as far as we know. It might never happen—or it might happen any day.

What will such first contact be like? I don't think we have to rely on our imagination alone to answer that question, for we can look back on history.

In the past, a given group of people had but limited knowledge of the world and might not even know of the existence of another group who lived a thousand miles away. There were occasions, therefore, when two such groups, until then altogether ignorant of each other, might meet. That, too, is a form of "first contact."

Generally, the meeting came about when one group was on the move, either fleeing disaster (of either human or environmental origin) or actively seeking new land. In doing so, it might come across another group, which was not in motion but living quietly at home. The moving group might be called the "contactors"; those at home, the "contactees."

In these cases, the contactors might be desperate enough for haven or property to attempt to enslave or kill off the contactees and seize their land. Even if the contactors were peaceful, the contactees, afraid of the newcomers and fearful of their motives, might attempt to block their entry, or to evict them. In either case, the result would be violence.

Ancient Egypt found itself flooded with Hyksos invaders from Asia in the seventeenth century B.C., and again by the Sea Peoples from the north Mediterranean coast in the twelfth century B.C. In both cases, the bewildered Egyptians didn't know who the invaders were or where they came from—they might just as well have come from Mars.

The Romans had to face a flood of Gauls from the north in 390 B.C., and the Cimbri and Teutons also from the north in 102 B.C. Again these barbarians loomed on the horizon with no warning whatever.

When people fight a known enemy, they are prepared and fight bravely enough. When the enemy is strange, unknown, and unexpected, the contactees suffer from surprise and possibly from a feeling that the invaders are, in some ways, demonic. The contactees are then very likely to suffer catastrophic defeat.

The Romans knew all about war and, in the third and fourth centuries, fought the Goths and Franks steadily. They were known enemies. However, when the Huns swept westward from central Asia in the fifth century, an element of panic was added at the sight of these strange, short, bowlegged, slant-eyed people, who seemed glued to their hardy ponies. The contactees were all but paralyzed. The process was repeated with even greater force in the thirteenth century, when the Mongols swept in from the east and smashed the quailing Europeans.

Nevertheless, the greatest of first contacts came in the fifteenth century and thereafter. Now it was the Europeans who were the contactors, and the peoples of all other continents who were the contactees. Everywhere the Europeans enforced their rule, enslaving or otherwise dominating the contactees, or "natives."

In that case, won't first contact in the science-fictional sense also mean war and slaughter? Or are there other factors involved?

Contactors from another planet (whether they are Earthmen who go there or aliens who come here) will surely be few in number in a possibly hostile world and would not dare offer violence unless they were enormously ahead in technology. And even if they were, they would be faced with totally different forms of life and intelligence, and they might feel it would be much more useful to study the new life-forms than to steal their world. Even if, for some reason, the contactors were violent and victorious, there might come pangs of conscience later on. Here on Earth many people of European descent are bitterly ashamed of what their ancestors have done.

What about the contactees? Here on Earth, the contactees generally did not offer violence until they were mistreated. They tended to greet the European newcomers peaceably enough, but they objected to having their land taken away and their people killed.

If in interplanetary first contact there is utter revulsion at the shapes and ways of the invaders, there may be instant violence from the contactees, even if they are in no way mistreated. The possibilities are various enough to give science-fiction writers much latitude in their consideration of first contact.

# 55

# Welcome, Stranger!

Human beings have never yet made an alien contact. To be sure, explorers, particularly Europeans in the great "Age of Exploration," have encountered strange organisms they had never seen or imagined before. In Africa, they discovered giraffes, for instance, and gorillas, ostriches, and, even as late as 1900, okapis. In North America they found bison by the millions, and moose; in South America, llamas and spider monkeys; in Australia, kangaroos, koalas, and kiwis.

None of these represented serious dangers. Human beings could deal with them rather easily. (Mosquitoes, lice, and tsetse flies were disease-spreaders that were much harder to deal with, but even these could be fought.)

By science-fiction convention, however, an "alien contact" is one with organisms that are equal or even superior to human beings in *intelligence.*

Such contacts have been made in myth and legend, where human beings have encountered gods, angels, devils, demons, afreets, genii, ogres, giants, and so on almost ad infinitum. Rational people are quite certain, however, that none of these stories represent the literal truth.

At worst, they are the creation of a human imagination intent on telling an interesting story. At best, they are dramatic distortions of something that exists. (Thus, a centaur may have originated when a culture innocent of horses encountered their first mounted raiders; the Scylla of Odysseus and the hydra of Hercules may have been exaggerations of the octopus, as Medusa may have been; dragons may have been combinations of snakes and crocodiles; and so on.)

Even the Bible tells how the Israelites, after wandering for forty years in the wilderness, reached Canaan and encountered giants there. Only fundamentalists, however, accept that as literally true. It is quite evident to rational readers that the Israelites were using the term metaphorically to speak of people not of giant size but of giant technologies, people who were capable of building walled cities and making use of chariots.

Nowadays, it has been suggested that human beings encountered extraterrestrial organisms in the form of intelligent "ancient astronauts" in places such as ancient Egypt and in pre-Columbian South America. It has also been suggested that people are constantly encountering intelligent aliens that arrive on Earth in unidentified flying objects (UFOs). Such suggestions are accepted seriously only by cultists and by unsophisticated people who are also quite ready to believe in Santa Claus and the Easter Bunny.

The encounters that come nearest to the science-fictional concept of "alien contacts" have been made between human beings and other human beings where each group had no knowledge, or even suspicion, of the existence of the other.

In ancient times, people generally knew details only of the people who lived in their own area. Any land beyond their constricted horizons, if it existed at all, might be a lifeless desert or might contain incredible monsters for all they knew. Settled agricultural people rarely ventured far from their farms, for obvious reasons.

Nomadic peoples, however, generally carried their possessions with them and, under conditions where overpopulation or bad weather decreased their chances for finding an adequate food supply, went on far-ranging expeditions looking for new land and more food. In that case, they were liable to stumble upon farming communities. Sometimes the nomads had the advantage of greater mobility and hardihood, and sometimes the farmers had the advantage of greater population and better organization and weapons. In either case one side or the other would suffer badly.

The immobile farmers were usually the more surprised, however, for the nomads in their wandering life easily gained the impression that the Earth was filled with a wide variety of people with strange customs and cultures. The farmers knew only their own, and invading nomads, especially if their intent was rapine and conquest, were greeted with the utmost horror and were often considered as monsters. The "Uighurs," who settled down to become Hungarians, are still remembered in modern fairy tales as "ogres."

Europeans experienced this a number of times in their history. In the sixth century, the Huns poured in from the east and reached central France before being stopped. In the ninth century, Vikings poured down from the north and harried all the coasts of Europe. Worst of all, though briefest, there was the Tatar incursion, again from the east, in the thirteenth century. This would not have been stopped at all if the Tatar Khan had not died in 1241 and the all-conquering horsemen had not been compelled to return to central Asia to elect his successor. The horrified Europeans distorted their name to "Tartars"—that is, creatures from Tartarus, the ancient Greek version of Hell.

It was the Europeans themselves who inflicted such semi-alien contact upon other people in the worst form. From the fifteenth century onward, European vessels, equipped with the mariner's compass and cannon, explored and dominated all the coasts of the world. They enslaved Africans and exploited Asians. The worst of all was when they took over the American continents, utterly destroying the Aztec and Incan civilizations and practicing genocide on a large scale.

One can well imagine that to native Americans and to Australian aborigines, who had been isolated beyond ocean barriers for twenty-five thousand years or more, the sudden arrival of pale-skinned strangers with a clearly superior technology must have seemed as horrifyingly strange as an invasion of Martians would to us.

These semi-alien contacts over the four-century span from 1500 to 1900 were long-continued assaults of Europeans on non-Europeans, and they firmly fixed in the modern mind that violence and merciless cruelty were the only forms of interaction possible.

Science-fictional accounts of true alien-contacts (human beings and nonhuman beings, both intelligent and both technologically advanced) had always been pictured as peaceful till the end of the nineteenth century. Such alien-contact stories were written as satires or as exploration stories in which civilized understanding prevailed among the intelligences. (Consider *Gulliver's Travels.*) It was not till 1898 that H. G. Wells published *The War of the Worlds* and the modern "alien contact" story was born.

The European nations had just finished carving up Africa (with Great Britain taking the lion's share) and, in the process, displaying a total, callous disregard for the "natives." Wells, with bitter irony, pictured the Martians fleeing their own dying planet to take over the Earth (Great Britain in particular) with similar callous disregard for Earth's natives.

And yet *must* this be so? The European view was poisoned by the fact that they felt themselves to have the only "true" religion, and the various peoples they met, with other religions, were therefore considered subhuman. Surely we know better now, and we need no longer behave like Spanish conquistadores or Puritan zealots when we meet other intelligences. And perhaps *they* need not either.

Have we (and they, I hope) learned that there is room for variety in the Universe and that we can gain more by learning from each other than by killing each other?

Perhaps.

# 56

# The Lost City

The time is 1500 B.C. The place is the island of Thera in the Aegean Sea, sixty-five miles north of the island of Crete. Few Americans have ever heard of it.

Because the city-state of Venice controlled the Aegean and parts of Greece during a portion of the Middle Ages, there are alternate Italian names for some places. Thera is also known as Santorini, therefore. Few Americans have ever heard of that, either.

Thera was a round island with a peak in the center. In 1500 B.C. there was a city on it, very advanced for its time, rich, prosperous, a great trading port. It had close connections with Crete, which was at the height of *its* civilization. Crete was the first maritime civilization, owned the first navy, and had cities without walls, for it relied on its fleet for defense. It had elaborate buildings, a cultured people, and (miracle of miracles) indoor plumbing.

Crete, with its Minoan civilization (for Minos, the ancient king of Crete in Greek legend), had existed for fifteen hundred years, traded with Egypt on an equal basis, and controlled much of the Greek mainland. Thera shared in its power and prosperity.

There was a catch. That mountain on Thera was not an ordinary mountain. It was an "extinct volcano." An extinct volcano is a good thing. It doesn't erupt, and its age-old lava makes for fertile soil. An extinct volcano is also a bad thing. It sometimes isn't entirely "extinct."

A volcano like Mauna Loa in Hawaii is only moderately dangerous. It has open vents inside, out of which lava is always heaving—and sometimes overflowing gently and slowly. It can sometimes destroy fields and habitations but people can get out of the way.

In an extinct volcano, magma slowly wells up from deep below and exerts a slow pressure against the long-solidified plug that blocks its vent. If it has its base on the sea, there might be a leak through which cold water can slowly penetrate. As it reaches the molten rock of the magma it steams, and the steam adds to the pressure until eventually, without warning, the rock plug gives way, half the mountain is blasted into the stratosphere, and there is an explosion that puts even modern efforts with nuclear bombs (if you don't count radioactivity) to shame.

That is what happened in 1883 to Krakatoa, a small island between Java and Sumatra. The explosion was called the loudest sound ever heard, for the atmospheric waves circled the world; the tsunami ("tidal wave") was felt half a world away; and three thousand people were killed by the explosion, ash, and water.

But Krakatoa shrank into something minor compared to the explosion at Thera in 1500 B.C. and then again, even worse, some decades later. Thera was destroyed in a moment in that first explosion; its civilization was wiped out; its people were killed. The island itself was torn apart, and the sea rolled over where it had existed. What is left now are some arcs, marking the rim of the island, and a small peak in the center, as the volcano grows again.

The first explosion virtually destroyed western Crete, but the second explosion got the whole island. The Cretan civilization was wiped out, and it was not long afterward that the Greeks conquered and occupied it so that never again, *never* would it play an important part in world history.

In fact, the whole eastern Mediterranean world was thrown into turmoil, as settled societies collapsed, and invading rovers, fleeing their own lands, grabbed what they could. Egypt held its own, just barely, but it was not a great power again until medieval times. The Hittite Empire was destroyed, and successive waves of invaders took over Asia Minor. The Philistines and Israelites invaded Canaan (a small thing, but magnified by the Bible). It took centuries before the civilized world settled down again.

The world of 1500 B.C. was not yet very literate, and there was no one in that troubled time who wrote up a careful description of what happened. At least there was nothing that survived very long.

However, it was too great a cataclysm not to remain in memory. The Egyptians, particularly, must have remembered. When matters are not written down in an authoritative and respected way, however, people have the privilege of distortion and embroidery and—people being people—full advantage is taken.

A thousand years later, the inquisitive and all-curious Greeks sopped up the tales fed them by the Egyptian priests, who had no objection to magnifying them for the benefit of people they no doubt considererd pushy barbarians.

And eventually, by the fourth century B.C., the tale reached Plato, one of the greatest minds and writers the world has seen. He told the event, not as sober history, but as a cautionary tale intended to teach. The tale was never finished.

It is clear, nevertheless, that he magnified it, either because that was how he received it or because he wanted to make it more effective. He wrote of a civilized island that was destroyed in a day and sank beneath the sea, but it was not just a small island in the Aegean Sea. That was too mundane, and the Greeks would not have believed it. It had to be a very large island—indeed, a *continent*—and it had to be in the romantic distance, where all myths and legends ought to be. To the ancient Greeks, one scene of wonders lay beyond the Pillars of Hercules (the Strait of Gibraltar) in the misty dimness of the Atlantic. And Plato called the sunken island "Atlantis."

Somehow Atlantis caught the imagination of the world. What Plato meant as a cautionary tale, as a fable, as a myth, became sober history to many people. Atlantis became a high-tech primeval civilization from which all the ancient civilizations, from Egypt to Mexico, derived their knowledge. Or it became a mystic land of powerful magicians and necromancy. Everyone snatched at it, and it was not until just a couple of decades ago that the true tale of Thera was finally captured by archeologists, who dug deeply into what is left of the island. But we can be sure that such research won't stop the marvelous tales of Atlantis.

We have in *Atlantis,* which I helped edit, a collection of those tales, from the sober realism of Sprague de Camp and the romantic semirealism of Karen Anderson to the racing adventure of Edmond Hamilton and the morbid mysticism of Clark Ashton Smith. It is a broad sampling of the things that can be done with Atlantis and an example of how it stimulated the imagination in so many different ways.

But the cautionary tale remains. Atlantis (like Thera) was destroyed, unexpectedly and suddenly, at the height of its power, and there is always the feeling that it was the punishment for arrogance and overweening pride.

Do we know another, nearer civilization marked by arrogance and overweening pride? And is there a well-known way in which, unexpectedly and suddenly, at the height of its power, it, too, might be destroyed? And can we imagine some day in the galactic future other species on other worlds speaking of a whole planet that suddenly, in a day, was destroyed? And might alien archeologists someday probe the slightly radioactive debris of the world in order to determine what really happened and whether the whole thing might not be a fable?

# 57
# The Bitter End

Every religion that I know of always has tales of the beginning of Earth and of humanity and of human technology. Some god or other being forms the Earth and sky, molds humans into shape, and teaches them the rudiments of fire, metallurgy, agricultural, and so on. Quite often the gods themselves are described as having a beginning. Zeus was once a baby, for instance.

Christianity is rather different from most, in that its creative god is supposed to be eternal. That introduces a great many philosophical difficulties, but we can leave that for theologians to deal with.

It is not so common for religions to picture an end to Earth and humanity. To be sure, we are all reconciled to the invitability of an individual end for each of us. (We generally palliate the pain by believing in the "transmigration of souls," that we will have another life here on Earth; or by supposing there will be an ideal immortality in an imagined heaven.)

But what about an end to everything, to the whole ball of wax?

That, too, is sometimes imagined. The prophetic books of the Old Testament speak of a Day of Judgment in which God brings everything to an end, destroying the vast majority in horrible ways while saving the inconsiderable number who follow the prophet's teaching exactly.

The best-known end, at least in our culture, is contained in the final book of the New Testament, "The Revelation of St. John the Divine."

The first three chapters of the book are routine, but after that is a long, long description of all the endless disasters that are brought upon the Earth by a vengeful God—the sea turns to blood, the stars fall from the sky, and so on. I find it all rather tedious and unconvincing. In the end, though, a new heaven and a new earth come into being for the inconsiderable remnants who are saved.

A non-Christian ending of violence can be found in the Norse *Eddas,* but since the version we now have only dates back to about A.D. 1000, it may have been influenced by Revelation. It tells of "Ragnarok," in which the gods and their adversaries battle themselves to mutual destruction . . . and then a new heaven and a new earth come into being.

Science, too, supplies us with bitter ends—of an astronomical nature. Prior to the 1930s, there was the feeling that slowly the Sun would cool down and Earth would become a lifeless block of ice. Since the 1930s, we have realized that the Sun is actually heating up and that someday the Earth will become a lifeless cinder. Both scenarios, however, would take place billions of years in the future and would happen so slowly that we would have time to escape (if we haven't died out long before).

For faster bitter ends, there is the possibility of the Sun going nova or supernova, but we now know that the Sun is too isolated and too small for either possibility. In the 1980s we have come to face the possibility of a cosmic impact with an asteroid or comet that might wipe us out just like that, but this isn't very likely in the near future. And if the catastrophe holds off for just a few centuries, we may develop the capacity to ward off such an eventuality if it seems to loom ahead of us. All these possibilities have been dealt with in science fiction stories, of course.

Forty years ago, a new possibility came into view. Human technological power, with the development of nuclear bombs and bacteriological warfare, reached the point where we can destroy all humanity, or most of it, in a few hours to a few months. We may have a human-caused bitter end.

Almost as soon as the first nuclear bomb, used in anger, had dropped on Hiroshima, science fiction writers began to write post-Holocaust stories, far more grisly than anything the stodgy imagination of the writer of Revelation was able to produce.

These are always cautionary tales, the message of which is "Don't let the bombs drop!"

No new earth and heaven arises by magic in such stories, but is there any hope that perhaps, more slowly, the indomitable spirit of humanity will recreate something? Well, read *Through Darkest America* by Neal Barrett, Jr., but remember—

Don't let the bombs drop!

# 58

# The Tail Wags the Dog

One good way to work out the future is to study the past. I've always thought so. A science fiction writer who deals with a future political situation *must* consider the past. He or she has no other guide. Besides, there's no reason to suppose that human beings ten thousand years ago, or ten thousand years from now, have not been, or will not be, guided by the same emotions, irrationalities, and, on all-too-rare occasions, the same iron logic that rules them now.

Thus, when, a great many years ago, I started writing about the fall of the Galactic Empire, I felt that I had no better guide than Edward Gibbon, who had written *The Decline and Fall of the Roman Empire* (which I had read twice).

We are now on the point of establishing settlements in space—both on the Moon and in artificial worlds placed in orbit about the Earth. If one wants to describe a world of this sort, one can scarcely avoid thinking of the time when European nations established settlements on the shores of other continents. That was a task which, in the sixteenth and seventeenth centuries, offered the same degree of technological difficulty, considering the state of the art then, that the settlement of space does today.

What happened back in the ocean age of our past that can guide us in the space age that is about to come?

The most dramatic story of all involved the British colonies on the eastern shores of North America. Not only is it particularly dramatic, but it is the best-known to Americans because those who lived that story are our cultural and historical, if not necessarily biological, ancestors.

What happened was that the colonies revolted against the mother country and won their independence. They set up a federal government to which the individual colonies/states gave up some of their sovereign rights, spread out across the continent, and grew steadily more populous, strong, and technologically advanced. Eventually, they became the most powerful country in the world and, in particular, advanced from a suspicious but weaker adversary of Great Britain, to a friendly partner and ally, to a somewhat contemptuous master. And recently, as members of the British Labor Party bitterly remark, when Reagan said to Thatcher, "Jump!", Thatcher replied, "How high?"

This took two hundred years, but the pace of events tends to quicken with time. Can we imagine that space colonies can also break away from Earth, quickly grow more advanced technologically, and, particularly in view of their dominating position in the sky, in a very short time come to dominate the Earth far less benignly than the United States dominates Great Britain?

My own feeling is that this is not likely. I reason it out as follows:

The settlement of space is a task that is extraordinarily difficult, expensive, and risky. (The recent *Challenger* disaster and the problem of finding the funds with which to replace the shuttle and put through the modifications and improvements that we are committed to is a sad example of that.) I don't think that either the United States or the Soviet Union can carry through the proper settlement of space by themselves, especially if they continue to spend an enormous percent of their money, time, efforts, and emotion on preparations for possible war with each other.

Either, then, the United States and the Soviet Union learn to cooperate with each other (and with the rest of the world) in order to move out into space, or else humanity simply won't move into space in any practical way. They will merely convert near space into another theater of war and bring closer the destruction of everything.

However, I have no intention of trying to force my views on anyone else. For one thing, I may be completely wrong.

So, for an alternate view (and a very exciting and thrilling view, too), in which humanity moves into space in a big way and manages to continue its bad old habit of intrigue and war, read *The Man Who Pulled Down the Sky* by John Barnes.

# 59

# The Ifs of History

There have been many occasions when the fate of humanity seems to have hung on the outcome of a single event that might have fallen this way or that with equal probability. What if Lincoln had said, "I don't feel like going to the theater tonight. I have a headache." Or what if Gavril Princip's gun had misfired when he aimed it at Franz Ferdinand of Austria?

My own favorite "if" of history involves a scientific discovery. Leo Szilard was a Hungarian scientist who had been driven out of Europe by Hitler's anti-Semitic policies. He knew that uranium fission, recently discovered, might make a nuclear bomb possible, and he wanted to be sure Hitler didn't get it first. He labored to get scientists in the field to practice voluntary secrecy and keep their discoveries to themselves.

Then, he and a pair of fellow exiles, Eugene Wigner and Edward Teller, labored to get still another exile, Albert Einstein, to write a letter to President Franklin Roosevelt, urging him to set in motion a secret project to build a nuclear bomb before Hitler did. Szilard knew that only Einstein possessed enough weight to be persuasive.

The letter was sent in 1941, Roosevelt read it, and, late in the year, he finally signed a directive that set up what came to be known as the Manhattan Project.

Now, he signed it on a Saturday, and, our society being what it is, people are often reluctant to do anything on a weekend. I could imagine Roosevelt tossing his pen onto his desk on the particular Saturday and saying, with a touch of irritation, "The hell with it. Let's take it easy. I'll sign it first thing Monday." It would have been such a natural thing to do.

Except that he *did* sign it . . . on Saturday, December 6, 1941. If he had waited till Monday, he might never have signed it, for Sunday, December 7, 1941, was Pearl Harbor Day; and, after that, by the time things cooled down, the whole business about the Manhattan Project might have been one with the snows of yesteryear.

What would have happened? Would Germany have gotten the bomb first? Would World War II have ended without the bomb, and would the Soviet Union have got it first during the Cold War? Would no one ever have developed the bomb? You could write three different stories about three different consequences from this one little "if" of history—if Roosevelt had yawned and said, "I'll do it Monday."

It's not easy to write such an if-of-history story. One little change might give birth to another, and then still another, until a later period becomes radically, almost unimaginably different from what we now consider reality. Or else such a change might produce a difference that, through some kind of social inertia, manages to converge until a later period is reached that is almost identical with what we call reality, except for a few amusing—or ironical—changes.

Science fiction writers occasionally dare to overcome the difficulty. There are two examples I have remembered with love over the decades. One is L. Sprague de Camp's "The Wheels of If," which appeared in the October, 1940, *Unknown,* and which dealt with a world in which the Muslems had won the Battle of Tours and the Celtic Church had won out over the Roman Church in the British Isles. The other is Ward Moore's "Bring the Jubilee," which appeared in the November, 1952, *Magazine of Fantasy and Science Fiction* and presented a world in which the Confederacy had won the Battle of Gettysburg and had established its independence. The latter was particularly touching because characters in it would fantasize about what would have happened if the Union had won the battle and America had remained intact. What a Utopian world they imagined would have resulted!

Well, now we have another attempt at an elaborate if-of-history. What if Justinian's attempt at reestablishing the Roman Empire had *not* overstrained it? What if the Byzantine Empire had been able to hold off the Zoroastrians or Persians and if Islam had never arisen to destroy the latter and permanently cripple the former. Might Byzantium have then carried Graeco-Roman culture, intact and in full, into the future?

Read Harry Turtledove's imagined result, *Agent of Byzantium.*

# 60
# The Sorry Record

When two organisms overlap too closely in a single environmental niche, they compete. It may not be purposeful; the organisms may not have the kind of brains that will make anything at all purposeful; but they will compete just the same. They will try to use the same habitats and live on the same food, and it is very likely that one will prove a bit more efficient than the other. The stronger will beat off, damage, kill the weaker; the better hunter or forager will leave the poorer to starve.

It is one of the mechanisms of evolution, usually expressed by the cliche "survival of the fittest" (except that you define the "fittest" as the one who survives, so that you have a nice circular argument).

To get a bit closer to home: We don't know exactly what killed off the australopithecines—they lived in eastern and southern Africa for two million years—but it may well be that genus *Homo,* wittingly or unwittingly, helped.

And *Homo erectus* may have been done in, at least to some extent, by *Homo sapiens,* while the Neanderthal variety of the latter was in turn done in by the modern variety.

We can't put ourselves into the minds of *Homo erectus* or *Australopithecus africanus,* let alone into what might pass as the mind of *Tyrannosaurus rex,* but we know very well what our *own* minds are like. We have minds that make it possible for us to know what we are doing when we callously mistreat others who are very much like ourselves, and do you know what we do? We *rationalize* our cruelty, and justify ourselves, and even make ourselves sound moral and noble.

Here is the first example I know of. Immediately after the biblical Flood, Noah planted a vineyard, made wine, drank it, and got drunk. And his youngest son, Ham, the father of Canaan, didn't show the old man the proper respect. (The Bible doesn't go into detail.) Noah therefore said, "Cursed be Canaan; a servant of servants shall he be unto his brethren" (Gen. 9:25).

In the time of King David and King Solomon, the Israelites controlled all of Canaan and enslaved the Canaanites and put them to forced labor, not because the Israelites were a master race and did as master races always do. Not at all. They did it (they said) because of the biblical curse on Canaan (one that was undoubtedly inserted into the Bible after the fact).

Very well, then, that was ancient times, and people were primitive and knew no better.

However, in modern times it was suggested that Ham, the youngest son of Noah, was a black and the ancestor of all the blacks that have existed since. This, of course, is entirely wrong, for the Canaanites, if we go by linguistic divisions, were as Semitic as the Israelites, the Arameans, the Babylonians, and the Arabs. They were *not* blacks.

However, it suited the slavemasters of Europe and America to pretend that Ham was black, because that made black slavery a divine institution and placed the blacks under that same curse the Israelites had made use of three thousand years before. When preachers from the slave states said that the Bible enjoined black slavery, Noah's curse was what they referred to.

In fact, you don't have to refer to a particular biblical verse to make yourself sound moral and noble. After all, when you enslave blacks, you free them from their slavery to their vile superstitions, their false religions, their primitive way of life, and you introduce them to the benefits of Christianity and save their souls. Since their souls are worth infinitely more than everything else they possess or can possess, you are doing slaves an enormous favor by enslaving them, and you're earning for yourself kudos in heaven, and flights of angels will sing you to your rest for being a noble slaveowner. (If you think that slaveowners didn't use this argument to justify themselves, you are very naive.)

In fact, to slaveowners, slaves were always responsible for their own slavery. To Aristotle, that great Greek thinker, those people who weren't Greeks were slaves by nature. These "barbarians" (so-called because they didn't talk "people-talk" the way the Greeks did, but made uncouth incomprehensible sounds like "bar-bar"), being natural slaves, were naturally enslaved. You did them a favor, obviously, by letting them be what they naturally were.

The very word "slave" comes, I believe, from "Slav," since to the Romans and the Germans Slavs were slaves by nature.

It's not even just slavery. The German Nazis killed hosts of Jews, Poles, Russians, Gypsies, and others. Did they do it because they were bloodthirsty, ravening beasts? Not to hear *them* tell it. They were purifying the race and getting rid of disgusting subhumans for the benefit of true humanity. I'm sure they thoroughly expected the gratitude of all decent people for their noble deeds.

And we Americans? Well, there is a story that the Turkish sultan, Abdul Hamid II, a bloody and villainous tyrant, visited the United States once and was tackled over the matter of the Armenian massacres. In response, he looked about him calmly and said, "Where are your Indians?"

Yes indeed, we wiped them out. It was their land, but we didn't enslave them, we killed them. We killed them in defiance of treaties, we killed them when they tried to assert their legal rights under those treaties, and we killed them when they submitted and did not defend themselves. And we had no qualms about it. They were "savages," and we were doing God's work by ridding the Earth of them.

There is a (possibly apocryphal) story that after Custer's Last Stand (the Massacre of Little Big Horn—it's only a massacre when white men get killed) a Comanche chief was introduced to General Sheridan (a Northern hero of the Civil War). The Comanche said, "Me Toch-a-way. Me good Indian." To this General Sheridan is reported to have replied, "The only good Indians I ever saw were dead." A very nice genocidal remark.

The history of human cruelty is revolting enough, but the history of human justification thereof is infinitely more revolting. To know that, we need only read *A Different Flesh,* by Harry Turtledove.

# 61

# Cleverness

From the earliest days of civilization, two chief motifs have run through fiction.

One motif is that of the strong man, the hero, the person who solves problems and achieves victories by means of muscular superiority. That always has been a surefire way of winning an audience, who never tire of hearing or reading tales of Gilgamesh, of Hercules, of Achilles, of Rustem, of Chuchulain, of Sir Lancelot. Nor do we tire of such things today, for contemporary fiction gives us our fill of mindless musclemen, from Tarzan and Conan to Superman and Rambo.

Second to this, however, is the motif of the clever person, the shrewd and (perhaps) devious person who solves problems and achieves victories by means of mental superiority. Such are the tales of Odysseus (Ulysses) and Sinbad the Sailor. These stories are more human and offer us a higher ideal.

After all, what right have we to glory in muscular superiority, when Homo sapiens is not by any means supreme in strength? A mature chimpanzee, smaller than ourselves, nevertheless has a much stronger musculature. Tarzan may be able (in fiction) to defeat apes and lions with his bare hands, but even he has no choice but to make friends with elephants.

On the other hand, when it comes to cleverness, human beings bear the prize as compared with all other species, so that the supremely clever person is the true gem of creation.

Of course, one expects a clever person to win out over those who are more muscular and powerful than him- or herself. They may do this by clever lies, by indirection, even by treachery. Audiences, untroubled by considerations of morality and ethics, applaud. Thus, the great heroes Achilles, Ajax, and Diomed could not take Troy with all their martial strength, but Odysseus succeeded by the strategem of a wooden horse filled with soldiers that the Trojans were talked into bringing into their city. Ever since, the "Trojan horse" has been a byword for treachery, but Odysseus is admired for his cleverness. And, if it comes to that, few nations at war (or individuals facing difficulties) have refrained from treachery if that would help them to victory.

In the Greek myths, the god Hermes, when only a day old, stole Apollo's herd of cattle, cleverly sheathing their hooves so that they would yield no tracks to give him away. Apollo found them at last, but clever Hermes had meanwhile invented the lyre and traded it to Apollo for the cattle. Hermes was much admired for all this and made the god of thieves.

On a less exalted scale, Autolycus and Sisyphus were, in Greek legend, clever thieves who constantly tried to outwit each other, and the picaresque tales of their misdeeds were found entertaining by the Greeks. (Autolycus was the grandfather of Odysseus.)

Such sneakiness pops up even in the Bible. The thirtieth chapter of Genesis describes how the patriarch, Jacob, outwits his father-in-law, Laban, by what can only be called underhanded ways (and faulty genetics, too). Jacob had also driven a hard bargain with his naive brother, Esau (see the twenty-fifth chapter) and had tricked his father, Isaac, into giving him, the younger son, the elder's share of the inheritance (see the twenty-seventh chapter).

This sort of thing continues right into modern times. We still admire resourceful outlaws, from Robin Hood to Jesse James, especially if we can convince ourselves, against all the evidence, that they robbed the rich and gave to the poor. We like stories about successful thieves, like Raffles, provided we give them a thin veneer of righteousness. Again, we enjoy the "caper" movies in which a bunch of clever rascals plan to seal from a bank or from a museum (or some other faceless victim). Generally, the plan is frustrated at the end, but there is no question that the audience's sympathy is with the malefactors.

It was comparatively late in the game that it dawned upon people that cleverness could be used not just to commit crimes but to thwart them. The first example I can think of occurs in the apocryphal biblical Book of Susannah, which was written about 100 B.C. It tells the tale of two corrupt elders who, thwarted in their attempt to seduce the beautiful and virtuous Susannah, bear false witness against her, accusing her of adultery. A clever young man, Daniel, then questions the elders *separately* and shows that their tales do not match and that, therefore, they are lying. Such tales of clever virtue continued to be rare, however.

Why does virtue limp and lag so badly in its race with vice? We must realize that most societies through history have been terribly unjust, have favored the rich and the well-born, and have treated the poor like animals. One cannot expect the poor (who have the virtue of numbers, at least) to see it as wicked if one of their own by superior cleverness outwits and discomfits those rich and well-born who oppress him so cruelly and unjustly. Naturally, the "Thief of Bagdad" is the hero of his particular tale, and "Reynard the Fox," that clever, low-born rascal, makes fools of wolves, bears, and lions (which represent the aristocracy). The aristocracy may applaud the tales of heroes and knights, but the people want to hear of successful thieves and of little tailors who slay giants by trickery.

Occasionally, some societies arise that manage to give most people some feeling that they, even if poor and low-born, have a chance to obtain justice. There has to be a governmental assumption, in theory at least, that a society is governed by the rule of law, without fear or favor, for rich and for poor, for high and for low alike. It is only then that there seems any point to the general population in supporting the law, and only then that it became possible to acclaim the clever person who bends his or her talents to the thwarting of crime.

I would like to suggest that this may be why the "detective story," as opposed to the "crime story," came into prominence (at least in the western world) first in Great Britain and the United States.

It is this that makes the nineteenth century particularly interesting in the history of "stories of mystery and crime." In it you can see the slow turn from the clever and daring criminal to the clever and daring detective.

We get our first clear glimpse of the latter in the works of Edgar Allan Poe, who wrote three stories of "C. Auguste Dupin." Two of these (in my opinion) are long and quite tedious, but the third, "The Purloined Letter," is in every way a modern story, even though it was written in 1844. The point it makes may not startle you, if only because the story itself has become so famous and has been used as a springboard for so many other stories making use of some variant of Poe's idea. But, to my way of thinking, it is the first literary production that unquestionably raises the clever person who supports society above the clever person who fights it.

And a half century after Poe, there came Arthur Conan Doyle, who clinched the victory once and for all. He invented Sherlock Holmes, who, for all time since, has been the quintessential detective and, many maintain, the most famous fictional character of all time. ("The Red-Headed League" is viewed by some to be his best short story.)

The Sherlock Holmes stories, all by themselves, persuaded many that there is something admirable in supporting and stabilizing the structure of society. It has not stopped crime (probably nothing ever will), and it doesn't entirely wipe out our sneaking admiration for the one who dares to fight society—which, after all, is never entirely just—but it has probably done more to rally support for the machinery of justice than all the preaching by all the ministers and philosophers in the world.

And that is not a bad accomplishment for a variety of fiction often derided as something suited for an occasional idle hour (and nothing more).

# 62

# In Days of Old

There are some words that reek of romanticism, and "knight" is one of them. Yet its lineage is rather low. It is from the Anglo-Saxon "cniht," which meant "boy" or "attendant," someone who attended his master and waited upon his needs. The German homolog, "Knecht," still means "servant" today.

Of course, if it is the king we are talking about, his attendants were often fighting men, and in medieval times that meant someone who could afford a horse and armor, and that, in turn, meant an aristocrat.

In other languages, it is the horse that was stressed rather than the service. In ancient times, to ride a horse was the surest sign of aristocracy (a war-horse, of course, and not a plough-horse), just as driving a Cadillac or Mercedes (not a Chevrolet or Volkswagen) does the trick today.

In literary Latin, the word for "horse" is *equus,* but in soldier lingo a "horse" was *caballus* (equivalent in English to "nag," or "hack"). It was the latter that came to be used for "war-horse." In Spanish, *caballus* became *caballo;* in Italian, *cavallo;* and in French, *cheval.*

Consequently, a horseman was *caballarius* in Latin, *caballero* in Spanish, *cavaliere* in Italian, and *chevalier* in French. All were equivalent to the English "knight." If we want to speak of the whole body of knights, you might talk of the "knighthood" of England, but it is more common to turn to French (for Norman-French, at least, was the language of the English aristocracy from the twelfth to the fifteenth centuries) and speak of "chivalry." To behave like a knight—that is, with courtly manners—instead of with the boorish behavior of malapert peasant knaves and varlets, is to be "chivalrous."

Actually, however, the romantic glow that makes knights seem so wonderful is totally a matter of fiction. In actual fact, knights, presuming on their horses and armor, were arrogant and insufferable in their behavior, especially to people unarmed and on foot. In English we have another word for "knight," "cavalier" (usually used for those who fought for King Charles I), and we all know what "cavalier treatment" means.

Incidentally, I made use of the word "knave" a few lines back. This means "boy" or "attendant," and the German homolog, *Knabe,* means "boy" even today. As you see, "knave" and "knight," which are now treated as opposites, meant precisely the same to begin with. (The German word for "knight," by the way is *Ritter,* meaning "rider".)

Ever since 2000 B.C., aristocrats did not fight on foot in the way the peasant-scum were forced to. The Homeric heroes fought in chariots whenever they could, and the Greek and Roman aristocrats were in the "cavalry" (the Latin equivalent of the French/English "chivalry").

Nevertheless, until the end of ancient times the cavalry never served anything but a supporting role. They were important mainly because of their speed of progress. They could spy out the enemy and they could pursue an already broken and fleeing foe. The actual fighting, however, was done by the steady and disciplined "infantry," the Greek line of hoplites, the Macedonian phalanx, the Roman legion. (The very word "infantry" is akin to "infant" and is another word meaning "boy." The term is a measure of the contempt held for the foot-soldier by the aristocrats.)

The role of the cavalry changed with the invention of the metal stirrup by the nomads of central Asia some time in the early centuries of the Christian era. What a difference that made! Without a proper stirrup, the cavalryman was insecurely balanced on his horse. And if he used a spear too incautiously, he could be easily pulled or pushed off his mount. Under those conditions, horsemen were better off using arrows, as the Parthian cavalry did. With a good stirrup, on the other hand, the cavalryman could wedge his feet securely and place the full weight of himself and his horse behind the spear. No footman of the period could stand against that.

When the Goths were fleeing from the Huns in the fourth century, they did manage to borrow the Hunnish stirrup, and in 378 the Gothic horsemen demolished the Roman legions at the battle of Adrianople. The cavalry was then supreme for a thousand years, and the era of knighthood began.

Still, however much knights were idealized and heroicized in fiction, in actual life they were cruel, despotic, and ferocious in their treatment of the lower classes, and when they were finally and disgracefully defeated, we all cheered.

The time came when the lower classes learned to fight the horsemen by keeping them at a distance and skewering them. In this the lower classes were greatly aided by that inevitable accompaniment of arrogant aristocracy: invincible stupidity. The Flemish burgers learned how to use the long pike in a steady line (the rebirth of the Macedonian phalanx) and slaughtered the French horsemen at the Battle of Courtrai in 1302. The English longbowmen massacred French horsemen from a distance at the battles of Crecy (1346), Poitiers (1356), Agincourt (1415), and Villeneuve (1420). The Swiss spikemen demolished the Burgundian horsemen in 1477, and by then gunpowder had established itself, and knighthood was all over.

But we still remember it in a golden glow of romance and, most of all, in the Arthurian legend—the tales of King Arthur of Britain and his Knights of the Round Table. In fact, any time we speak of "knights" we think of those tales and, most of all, of Sir Lancelot.

The Arthurian legend began with Geoffrey of Monmouth, who wrote his *History of British Kings* around 1136 and in the process talked of Uther Pendragon, his son Arthur, and their helpful wizard, Merlin. It was not history but myth and legend, yet it fascinated its readers, who then, as today, would rather have history appeal to their superstitions and patriotism than to any abstract and bloodless passion for truth. If you want an excellent modern retelling of Geoffrey's tales, read *The High Kings,* by Joy Chant (Bantam Books, 1983).

About 1170, a French poet, Chrétien de Troyes, took up the tale and added straightforward romance. It was he who first invented the adulterous passion of Lancelot and Guenevere, and the mystical tale of the search for the Holy Grail. Inasmuch as Chrétian made no pretense to even the shadow of historical truth, his tales were even more popular than Geoffrey's.

Sir Thomas Malory put together the scattered fragments of the Arthurian legend into *Morte d'Arthur* (*The Death of Arthur*), and it is his version, published in 1485, that we know best today.

The legend has never died, and in each century it has been retold. In modern times there are Alfred, Lord Tennyson's *Idylls of the King* (1859), Mark Twain's *A Connecticut Yankee at King Arthur's Court* (1889), and T. H. White's *Once and Future King* (1958). From the last of these, the musical *Camelot* was taken. Most recently, there is Marion Zimmer Bradley's *The Mists of Avalon* (1982).

The Arthurian legend is strictly fantasy. It is loaded with wizards, enchantresses, spells, and magicianry. Those who attempted to remove the fantasy and present the legend in a realistic manner were least successful. I found Tennyson to be dish-waterishly dull, for instance. Twain introduces the time-travel motif, which makes for anachronistic amusement but by turning Merlin into a flim-flam faker, he greatly detracts from the interest of the tale.

White, on the other hand, especially in *The Sword in the Stone* (1939), the first volume of his tetralogy, *adds* to the fantasy, and his version is superior to Malory for that reason (in my opinion). The same can be said of Bradley's painstaking tour de force.

It is not surprising, then, that modern fantasy writers turn every now and then to knightly romanticism and, in particular, to aspects of the Arthurian legend and try their hand at it. Here in *Cosmic Knights,* which I helped edit, for instance, we have ten stories of different types, all united by their possession of knightly heroes.

Zelazny, for instance, adds a moving epilogue to the Arthurian legend. Young, on the other hand, repeats Twain's attempt to deal with it anachronistically and humorously, but he does so more successfully, I think. There are satires on the whole notion of knightliness, from the careful self-consistency of de Camp to the rollicking implausibility of Hunter and the delightful ingeniousness of Grahame. There is even the straight science fiction (however fanciful it may seem at first blush) of Anderson and the horror tale of Schweitzer.

I am quite certain you will enjoy this collection, even though you might never again be able to think of knights in quite the fashion you used to.

# 63

# Nonviolence

Forty-seven years ago, in the very first story of my *Foundation* series, I had my hero say, "Violence is the last refuge of the incompetent."

Not everyone thinks so, of course. Big men are often firm in the belief that violence is the correct way of deciding a dispute, especially if they are facing small men. People with guns are similarly firm in that belief, especially if facing unarmed people. And a mob believes in violence when facing an individual. In short, if violence seems as though it is on your side, you like it.

Of course, if big people, armed people, or a mob of people just happen to be on the side of the right, then violence becomes a way of ensuring justice, doesn't it? We watch an infinite number of movies and television plays in which right defeats wrong because it always happens that the "good guy" just happens to be bigger and stronger than the "bad guy", the good guy just happens to be able to shoot faster and more accurately than the bad guy, and so on.

Sometimes it is clear that the good guy is more stupid than the bad guy, who is very often presented as the most intelligent person in the drama.

However, the good guy is sometimes saved by the fact that he has a horse that is smarter than he is—and besides, the faster gun beats the faster mind every time. And so we are taught not only that violence is the path to right and justice but that intelligence is usually wicked.

My own feeling is that violence usually brings about the victory of the unjust and barbaric. As an example, the number of muggers who knife, shoot, or otherwise inflict violence on peaceful citizens is far greater than the number of muggers who are themselves beaten up by those same peaceful citizens.

The answer of the violence-lovers is: "Give all those peaceful citizens guns of their own."

Why not? In this way, we convert everyone to muggers of a sort, and we have gunfights with each side claiming the other drew his or her weapon first. (My own bet, in this connection, is that the more experienced gun-toter—i.e., the greater criminal—will win virtually every time.)

The situation of mugger-versus-mugger is nowhere truly to be found in any nation, but I am told that the nation that most nearly approaches it is the United States, for nowhere else is there such a tender love for guns and weapons.

If we want the real mugger-versus-mugger scenario, however, we must consider the international situation. Every nation that is larger than a vest-pocket is armed. Every nation is convinced that in all disputes it is right and the other nation is wrong. Every nation seems to feel that it is thoroughly justified in using all possible force any time it gets mad.

So what do we have? Terrorism and the constant never-lifting threat of nuclear war.

What a great world!

Isn't there any way of running the world without making violence the decisive way of judging between right and wrong? Says the lover of violence: "No way. We have always had wars. That's just the way things are. The answer is to make ourselves stronger than the other guy."

That is exactly what the other side says, too.

We consequently live in a world in which the Soviet Union disposes of enough weapons to kill everyone on Earth fifty times over, and feels insecure because the United States has enough to kill everyone sixty times over. Then, because we suspect that the Soviet Union really disposes of enough to kill everyone seventy times over, we feel insecure, too.

What a relief both sides would feel if they could believe with all their heart and soul that the enemy, no matter how hard they tried, could only kill everyone once—just once. We would all dance in the streets, wouldn't we?

Does this strike you as making sense, by the way? Does it sound like the action of competent people?

A reader once wrote: "If violence is the last refuge of the incompetent, why is there so much violence in the world?"

And I replied, "Because there is so much incompetence."

In any case, Judith Moffett's *Pennterra* takes up the matter of violence and nonviolence.

# 64

# Empires

The Latin term *Imperator* was used by the Romans as a title for the leader of an army. It is roughly equivalent to our "General." Eventually, it came to be applied, in particular, to the supreme leader of all the armies of Rome, to the *generalissimo,* so to speak.

By the first century, B.C., the Roman realm was more and more coming under the influence of a single man, with the Senate and the other government officials little better than puppets. The single ruler held his power from the fact that the army was loyal to him and would obey his commands. Therefore the most important and realistic title he held was Imperator, and from Augustus Caesor on we no longer speak of the Roman Republic, but of the Roman Imperium; or, to use the English distortion of the term, the Roman "Empire." What's more, the Imperator, by the same distortion, became "Emperor."

The Roman Empire arose from the gradual conquest of the entire Mediterranean world by the Romans, who originally ruled a small section of central Italy. Because of this, the term "Empire" came to refer to any group of different peoples (or cultures, or nations) ruled by a people (or culture, or nation) that has conquered or absorbed them all. Such an empire is usually ruled by a single person who is a member of the conquering unit, and in the empire the conquerors usually have special privileges.

The term has been extended to all such realms, both before and after Roman times.

The first Empire in history is usually considered to be the Akkadian Empire, established by Sargon of Agade about 2320 B.C. The smallest realm commonly given the name is perhaps the Athenian Empire, which, for about fifty years, ruled the shores of the Aegean Sea.

On the whole, empires cannot extend their rule without limit, since problems of communication and administration increase rapidly with increasing area. The Roman Empire itself reached its maximum extent about A.D. 125, at which time it was rather overextended. It moved to the defense and contracted very slowly over a period of thirteen centuries before the last scrap of it (the city of Trebizond, in Asia Minor) was submerged.

Still, as we move forward through history, from the time of Sargon of Agade onward, we find that technology slowly (but rather steadily) advances. With technological advance, the ability to conquer, defend, and administer an empire increases, so that, on the whole, the size and power of empires has increased with time.

Prior to modern times, the largest and most spectacular empire was that of the Mongols. Starting from almost nothing, Genghis Khan and his generals conquered a tract of land that is roughly marked out by the modern lands of the Soviet Union, China, Afghanistan, Iran, and Iraq. What's more, the colossal task was all done in fifty years. (However, the Mongol Empire broke apart, and most of it vanished, in another fifty years.)

In modern times, empires have grown larger still, especially since advances in the science of navigation have made it possible for a nation to control the coasts of distant continents, and to work one's way inward wherever weakness exists. The first noncontiguous Empire (a realm connected by sea-lanes rather than by a continuous stretch of land) was that of Portugal, established about 1500.

The largest, most populous, and most powerful empire of all time was of this kind. It was the British Empire, which reached its maximum extent after World War I, when it controlled roughly one-quarter of the land area and the population of the entire planet. The British Empire was still at its maximum extent in 1945 at the end of World War II. But, in the most amazing reversal in history, that empire vanished completely by twenty years after that date. The British, without military defeat, simply gave it up, as something they could no longer reasonably retain.

Nowadays, the age of old-fashioned empires is over. The large nations of today are either more-or-less homogeneous in culture (China, India, Indonesia) or they are "federations"—that is, unions of equal parts (at least in theory). Thus, the United States is a federation of states; Canada, of provinces; the Soviet Union, of socialist republics; and so on. Similarly, the League of Nations and the United Nations are examples of (very weak) federations of nations.

And where, in all this, does science fiction come in? Well, it is inevitable that science-fiction writers look forward to a world in which the human species continues to increase its range.

Starting from a small patch of land in east-central Africa, the various hominids, culminating in *Homo sapiens,* have spread out over all the Earth's surface. It seems inevitable to the ever-romantic mind of the science-fiction writer that we must now spread first to the Moon, then to the remainder of the Solar System, and finally to the stars.

And with that expansion, there is bound to be the notion of expanding political systems.

What kind? It would be pleasant if, along with the various advances of technology that *must* accompany (and, indeed, *precede*) expansion through space, there should be an equivalent evolution of political systems, administrative techniques, societal and economic devices.

Easier said than done. The empire remains the favorite symbol of the large state, and it has taken over the imagination of the science-fiction world. Nor is it the most recent empires, after the fashion of the British, that are used as models. Rather, the mood, the atmosphere, and the paraphernalia is that of ancient and medieval times.

For this, I, myself, am largely responsible. Before my time there were stories about Earthmen meeting other intelligences, or living on other worlds as well as Earth, of conquering or being conquered; but the first attempt to write a series of such stories with a rational historical background, and to achieve considerable popularity as a result, was mine. It was my *Foundation* series, which began appearing in 1942, that set the fashion.

In 1942, to be sure, the European empires, particularly the British, were still in existence, and they seemed likely to be eternal, but they didn't influence me. I modeled my "Galactic Empire" (a phrase I think I was the first to use) quite consciously on the Roman Empire.

Ever since then, other science fiction writers have been following the fashion, and have written series of their own after the fashion of the *Foundation* series. In fact, in the late 1970s the Galactic Empire reached the movies in the enormously popular *Star Wars,* which, here and there, offered rather more than a whiff of the *Foundation.* (No, I don't mind. Imitation is the sincerest form of flattery, and I certainly imitated Edward Gibbon, so I can scarcely object if someone imitates me.)

In this book, then, we have nine stories by nine authors illustrating nine different versions of Galactic Imperial history, since each is part of a series of at least three stories. One story included is, inevitably, one of mine, and it happens to be one that is the least typical of my Empire stories.

I hope that in reading the stories in this anthology you will enjoy comparing and contrasting the manner in which top-notch writers use the Imperial theme to consider problems that, in sheer size at least, transcend those that would have involved merely planetary empires.

# 65
# The Last Man on Earth

What is the oldest surviving last-man-on-Earth story?

Well, it's almost as old as history. History begins with writing, for it is only through writing that we can get a connected story of events. We can deduce things from various nonwritten artifacts—pottery, paintings, jewelry, tools—but that's not the same thing. What we deduce from such items is "prehistory."

History, then, begins about 3100 B.C., in the land that is now called Iraq. Along with the lower course of the Tigris and Euphrate Rivers there lived a people we call Sumerians, who were the first to invent writing, as well as various other things. (They were very clever people, obviously.)

As is monotonously true of all peoples, the Sumerians had to deal with natural disasters, and about 2800 B.C., they had a really bad one. The rivers, which were their source of life, thanks to irrigation and agriculture, overflowed. They did this periodically, as rivers will, but this time they *overflowed.*

Why the incident was so unusually disastrous—whether because of particularly heavy rains, particularly high tides, a tsunami, a meteor splashing into the Persian Gulf—we don't know. In any case, much of the valley seems to have been flooded, and there must have been great loss of life.

Eventually, the flood receded, and there was a recovery, but forever after the Sumerians dated everything as having happened "before the Flood" or "after the Flood." Naturally, since they didn't know what caused the flood anymore than we do, they blamed the gods. (That's the advantage of religion. You're never short an explanation for anything.)

There was a science-fiction writer in Sumeria who got the brilliant idea of telling the story of the Flood, but of wringing a little more drama out of it than had actually existed. Why just drown thousands? Drown everybody! Well, except for one person and his family, so that he could account for the fact that there were still human beings on Earth.

The story grew (I dare say other writers added their own bits), and eventually there came into existence the *Epic of Gilgamesh, King of Uruk.* As part of the tale, the writer described how and why the gods decreed a flood that would (and did) drown *the whole Earth.* (In the first place, most of the Sumerians probably thought Sumeria and the immediately surrounding nations *were* the whole Earth, and in the second, did you ever know a writer—except me, of course—who could resist embroidering the facts?)

He then described how *one* man—just *one*—managed to escape by the kindness and guidance of one of the gods. The man was Ut-Napishtim, and he was the first person we know of, by name, who was "the last man on Earth" in a science-fiction story.

The tale of Gilgamesh was extraordinarily popular, and it was still to be found in the library of Asshurbanipal, the last great king of Assyria, over two thousand years after it was written. (We found it in the ruins of that library over two thousand years after it was destroyed.) It undoubtedly spread through the entire ancient world and helped inspire other "strong men" stories: Hercules among the Greeks, Samson among the Israelites, Rustem among the Persians, and so on.

What's more, the story of the Flood was seized upon. In the first place, it was very dramatic; in the second place, it was considered actual history; and in the third place, most cultivated people lived along rivers or coasts and were familiar with floods.

The result was that the Greeks had their story of Deucalian and the Israelites their story of Noah. The Israelites picked up much of the material in the first eleven chapters of Genesis during the period of the Babylonian captivity in the seventh century B.C., and that included the story of the Flood. Except for getting rid of the polytheistic bits, they stayed pretty close to the details as given in Gilgamesh. (In our cynical modern world, we would call it plagiarism.)

There are two things I want to note about this first last-man-story: One, Noah wasn't really the last man—he had three sons with him; in the second place, there were also four women present—his wife and the wives of his three sons. The women weren't important, though; the Bible doesn't even bother to give their names. (In the Greek tale, Deucalian has a wife, too, and although the Greeks weren't great shakes at women's rights—quite the reverse—they at least named her. It was Pyrrha, if you're curious.)

Even more important, the biblical version of the tale of Gilgamesh was accepted as sober history for thousands of years. It wasn't until about 1800 that geologists began to realize that there had never been a world-wide flood on Earth. Yet even *today* there are many who are certain there was indeed a world-wide flood, "because the Bible says so." This includes the "creationists," who are very anxious to teach their version of Babylonian mythology in the schools as "science," by George—so don't tell *me* that science fiction writers don't have influence!

There is such a thing as progress, however. We have come a long way since the time we had to drag in some god throwing a tantrum to account for humanity being reduced to a last fragment.

It was only a few years ago that scientists described a natural disaster that may really have happened and may very nearly have sterilized the Earth: the impact of a sizable asteroid. It did (many say) succeed in wiping out the dinosaurs, and such a thing may have happened half a dozen times before in Earth's history.

There are also vast pandemics, such as the Black Death, which is supposed to have wiped out a third of the human species in a third of a century. There are imaginative exercises of more modern science-fictional views than were available to the Sumerians: time travel to the far future, invasions of conquering extraterrestrial hordes, and so on. Most of all, there are the prospects of a deadly nuclear war, which seems the most likely of all paths to last-mandom, if paths there must be.

So here in *The Last Man on Earth,* which I helped edit, we ring eighteen changes on the theme—eighteen successors, in a manner of speaking, to Ut-Napishtim. Just for fun, see if you can think up a nineteenth.

# 66

# Image of One's Self

When I was very young, I read the myth of Narcissus. Narcissus was a handsome youth who rejected all the young women who were dying for his embraces. One of them prayed he, too, would feel the pangs of unrequited love; so, of course (since the gods are always eager to answer unkind prayers), he did.

He spied his reflection in the water, thought it was another youth, fell in love with its beauty, attempted futilely to embrace it, and finally drowned when he tried too hard. The myth has made such an impression on people that the word "narcissism" is a recognized psychiatric term for morbid self-love.

I did not like the myth at all. To my childish self, it seemed incredibly stupid. How could Narcissus mistake his own image for another youth? (I was also more than a little puzzled that he should fall in love with another youth, rather than with a maid, but the problem of image overrode that.)

No one bothered to explain that point to me. They were only interested in explaining the moral: If you are unkind to others, others will be unkind to you; if you are too fond of yourself, you will find life unpleasant.

I saw the moral, of course; it was obvious. What I wanted, though, was a technological explanation, and I never got it. I had to work it out for myself as I grew older.

The point is that in primitive times it was perfectly possible for one to see the faces of those around one with perfect clarity (assuming eyesight was normal). One could tell, at sight, the identity of every human being with whom one was acquainted. One could also tell, at a glance, that some person one encountered was a stranger, someone he or she had never seen before.

There was one exception. Under primitive conditions, no person could see his or her *own* face. If, through some form of magic, his or her own face were presented, one would have no choice but to consider it that of a stranger.

To see your own face without magic, you need a smooth, reflecting surface. A piece of smooth, unflawed glass will do, rather dimly, for it lets pass much more light than it reflects. A piece of glass backed by a smooth layer of metal will do it with near-perfect efficiency. That would be a "mirror" or a "looking-glass" (for what glass do we look at more eagerly than at a mirror?). In primitive times, however smooth unflawed glass, with or without metal backing, was unavailable.

One could simply polish a flat piece of metal. That would do well enough until it tarnished. In primitive times, though, a piece of polished metal large enough to see your face in was not an easy thing to get.

That left the surface of water, which was usually so broken up by waves, foam, and (even in quiet ponds) ripples that a reflected face was too disturbed to make much impression. If, then, Narcissus had come upon a pool *so* quiet that he could see a clear impression of his own image, you can well imagine that it was the *first* time he had seen it. The image would seem to have belonged to a strange youth, one who was hiding under the water.

Once you understand the tale of Narcissus in its true light, you may come across a sudden analogy. Human beings try to understand the Universe little by little. They look at this aspect and at that aspect and learn to analyze the appearance and characteristics and begin to understand it. We can be very proud of the fact that, all through our existence, our understanding of the Universe has grown enormously. But, then, why not? We are looking at every aspect of the Universe with one chief tool, the human brain (and the intelligence with which it is associated). Since the human brain is by far the most complexly interrelated piece of matter that we know of, we are using a complex tool to understand the much less complex objects that we are observing. Given enough time and thought, we *must* understand.

But then comes a point when we wish to contemplate the human brain itself (and the intelligence with which it is associated). Now we are attempting to understand something extremely complex by making use of a tool that is no more complex than it is. The situation is analogous to that of someone using his eyes to see his own eyes. As soon as we face the problem of the human brain and human intelligence, we are in Narcissus's case—faced with our own image, and therefore doomed, perhaps, to misunderstanding and death.

But there are solutions. Narcissus's solution would have been familiarity with a mirror.

Our solution is that we are not using *a* human brain to study the human brain and human intelligence: we are using *many* human brains to do so.

It is not *a* scientist who is studying the human brain; it is, rather, the community of science. This community has a complex structure of its own, with published papers, with frequent conferences, with communications in which different thinkers present different pictures, different interpretations, different observations. In kaleidoscopic fashion, these all melt together and grow almost without the volition of any individual, so that understanding increases at a speed and to an extent that any one person would find amazing.

What we (who are intelligent but who have done no work on intelligence) need is someone who has followed the work being done on all the aspects of science that impinge on intelligence and present them to us in orderly fashion and, if possible, with his or her own thoughts and ideas added to the mix.

This is precisely what Victor Serebriakoff has done in *The Future of Intelligence.* Himself a person of monstrous intelligence, he has obviously read, studied, and thought about every aspect of human intelligence, and here it is for us to share with him. He comes to rescue us from Narcissus's fate.

# 67
# Psychology

The word *psyche,* in Greek, referred originally to "breath," which, of course, the Greeks did not understand in the modern, scientific sense. To them, the breath was something ethereal and insubstantial, something that was, somehow, intimately related to life. Stones do not breathe, nor do dead human beings.

The word came to be translated into the English word "soul," which is also viewed as something ethereal and insubstantial and somehow intimately related to life. A more exact definition, however, loses itself in theological subtleties and uncertainties.

If we are to define "psyche" or "soul" without reference to theology, we might think of it as the inner core of being that the physical body houses. It is the personality, the individuality, the thing you think of when you say "I." It is the thing that remains intact and whole, even though a limb is lost, eyes are blinded, or the body itself is ill, wounded, or dying.

"Psychology," then, is the systematic study of that inner core that is you, and the word we are most apt to use to represent this in these nontheological times is not "soul," but "mind." Psychology is the study of the mind.

Psychology is fascinating in that it seems to exist at the end of two extremes of knowledge. In some ways, everyone understands it; in other ways, no one does. Other sciences can share these extremes of epistemological character—perhaps all sciences do—but surely none to the extent that psychology does.

For instance, to understand why a billiard ball behaves as it does, why it moves when struck, how it collides and rebounds with the cushion or with another billiard ball, how speed and direction alter as a result of collision—all this requires a good knowledge of the principles of that branch of physics known as mechanics. In reverse, the principles of mechanics can be worked out by a careful study of the minutiae of billiard-ball behavior.

And yet those who are expert at the art of billiards have not necessarily studied physics and mechanics, might never have heard of the conservation of momentum, might not appreciate the mathematical complexities of angular momentum produced by the placing of "English" on balls. Yet they make billiard balls do everything but cook dinner, and they do it by meticulous attention to principles they do not know they know.

The same may be said of those who pitch baseballs with complex virtuosity, and batters who strike at those pitches with an artistry of timing. They may earn millions for their mastery of the applied science of mechanics even when (conceivably) they have never learned the simplest fundamentals of physical science.

You can understand the laws of science in a very useful sense merely through careful observation and practice, for science is an organized system of describing the real world, and *you live in the real world.* You can't help but learn to describe the world merely by virtue of that fact, even if your description is not in the conventional terms that scientists have worked out and agreed among themselves to use.

It is not surprising, then, that there are people who have come to understand the human mind well through having observed others, through living and interacting with them, through becoming aware of their habits, responses, and peculiarities. No one can read Shakespeare, Dostoevski,

Tolstoy, Dickens, Austin, Moliere, Goethe, and any of innumerable others, without seeing that each has a deep understanding of humanity in all its varieties and perplexities, even though none of them studied psychology in any formal way.

This nonscientific understanding of psychology is undoubtedly more widespread than that of any other science. Sportsmen may deal unwittingly with physics, cooks with chemistry, gardeners with biology, sailors with meteorology, artists with mathematics—yet these are all specialized occupations.

*Everyone,* however, without exception, must deal with people. Even recluses must deal with themselves, and this might be enough, for each of us may well have within himself, or herself, all the virtues and vices, brilliance and foibles, aversions and tendencies, of humanity in general.

Therefore we may conclude that in some ways psychology is the best and most widely understood science.

And yet . . .

The human mind, born, as it must be, of the human brain, is an extraordinarily complicated thing. The human brain is, with little doubt, the most complicated, subtly interrelated lump of matter we know of (with the dubiously possible exception of the brain of the dolphin—which is both larger and more convoluted than the human brain.)

In studying something as superlatively complex as the human brain, we should naturally expect ourselves to be frequently at a loss. That is all the more obvious when we stop to think that we study the human brain with nothing more than the human brain. We are asking complexity to comprehend equal complexity.

It is no wonder, then, that even though billions of human beings have, throughout the history of *Homo sapiens,* been studying themselves and others in a casual and unsystematic way, and even though extraordinary geniuses have illuminated the human condition in literature, art, philosophy, and, in these latter days, science—vast stretches of the uncertain and unknown remain (and more so, we can be sure, than in any other science). Even those areas most studied and expounded are bound to remain, to some degree, in dispute.

And so, in some ways, psychology is the least understood science.

Consider, too, that the solution to all the problems that press, and have pressed, upon humanity through all its history rest, to a large extent, on the misfunctioning of the human mind. Some problems can seem totally independent of us, and intractable to any human effort—the coming of an ice age, for instance, or the explosion of the sun—and yet even there the human mind might conceivably foresee the event and choose actions that will ameliorate the effect, even if only by making death easier. Good will, reason, and ingenuity are needed (and are often lacking).

On the other hand, human folly (or, at the least, insufficient wisdom) present us with ever present and increasing danger. If we destroy ourselves through nuclear war, or overpopulation, or waste of resources, or pollution, or violence and alienation, then part (perhaps most) of the cause will rest with the inability of our minds to recognize the nature of the danger, and the reluctance of our minds to accept the necessity of taking those actions required to avert or ameliorate that danger.

There is no question, then, that psychology is the most important of the sciences. We can live, however primitively, with very little knowledge of any or all the other sciences, but let us not understand psychology and we are surely lost.

What role does science fiction play in all this?

Science fiction writers are not, on the whole, better or more understanding than other writers are, and there is no reason to look to them, as individuals, for a better illumination of the human condition.

In science fiction, however, human beings are pictured as facing unusual situations, bizarre societies, unorthodox problems. The effort to imagine the human response to such things can cast a light into the shadows in a new way, allowing us to see what had not been clear before.

The stories in *Hallucination Orbit* have been selected with that in mind, and each is preceded by a special headnote written by my coeditor, Charles Waugh, who happens to be a professional psychologist.

# 68

# Show Business

There is a story about an unfortunate who was sitting at a table in a bar, huddling over a beer, and wearing a look of intense misery on his face. A newcomer, observing this, and feeling his soft heart ache at the sight of unhappiness so obvious, took his drink to the table and said, "My friend, you seem unhappy. Would it help to talk about it?"

The unfortunate heaved a sigh and said, "It is just that I have no friends. Even here in this bar I am avoided. It is kind of you to come here, but I'm sure you will leave soon."

"Why should I?"

"Well, by profession I am one of the elephant-tenders at the circus. It is my job to sweep up after the elephants, and, as you can well imagine, the odor of elephant dung clings to me."

"Yes," said the other, uneasily. "I am aware of a certain effluvium."

"There you are. No matter how I wash or how often I change my clothes, there is this distressing odor that drives everyone from me and condemns me to a life of loneliness and despair."

Said the other, consolingly, "But, sir, why do you not find a different job?"

The unfortunate's eyes widened and he sprang to his feet. "What," he thundered, *"and leave show business?"*

* * *

We laugh at the joke, but we must see that there is a germ of truth to it.

There are not many professions which are, in and of themselves, glamorous; but surely of the few that are, show business, in all its many manifestations, ranks at the top. Consider the reasons:

1. Show business is a thing of tinsel and pretense. Butchers are only butchers and clerks are only clerks. An actor, however, can be a king or an angel or a murderer or a businessman or—anything at all. And he or she can switch from one to another from one night to the next. Naturally, we all know that these are imaginary identifications; that underneath the costume, inside the paint, aside from the pear-shaped tones, there is just a human being who is, perhaps, in some ways, far less impressive than the local butcher, both physically and morally. It doesn't matter; the appurtenances cling, and we envy the excitement and glamor of the surface life we perceive and care nothing for the dull anticlimax it may hide.

2. Show business lures the practitioner, as well as the observer. To be in show business is to be in one of the few professions (if not the only one) in which one's service produces an instant reward. A butcher might possibly be thanked, eventually; a writer might even be adulated for a best-selling book he or she has written, eventually; but someone on the stage, having delivered a clangorous line, or performed an unusual feat, or perhaps merely making an entrance, is greeted at once by the thunder of applause, and he or she knows it is meant for him (or her), and for him only. One takes one's bow in pleased acknowledgment, and, at that moment, one is content with one's role in life. Even though one might be underpaid, even though the dressing room and everything and everyone else one must work with are mean and poor, even though acting is a profession that so many super-respectables consider a kind of vagabondage, that applause and that bow is, for the while, all one expects of life.

3. Show business can be profitable. Someone in show business can become supremely wealthy, sometimes as a result of a single success. This should not hide the fact that the large majority of people in the profession

make only moderate livings, often very poor ones—but it does. It is the wild successes that attract attention. This is, to some extent, true in every profession. The very occasional billionaire executive obscures the thousands who achieve no more than middle-class suburbia. The great best-selling writer hides the many thousands of writers who are acquainted chiefly with rejection slips and whose novels are left at the post after being published. In no profession but show business, however, are the anomalies of success so clearly on view, so sharply in focus. In no other profession do they so completely and showily wipe out all that is mediocre or poor—or even merely modestly proficient.

4. In other professions where success can be noted and admired, the admiration is usually for money or power. One can note the great oil billionaires and envy them their wealth but nothing else. One can observe the President of the Unted States and envy him his ability to command the headlines, but not wish to be in his place for any other reason. In show business, however, the great successes are very often those whose faces and bodies are what we conceive to be virtual symbols of beauty and of sexual attractiveness. We, the admiring public, want not only their wealth and power; we want to *be* them, or, at the very least, to *look* like them. We may envy a billionaire or a president, but we *adulate* a movie star. We form fan clubs, we scream for them, we follow them, we fight for an autograph, and we dream of them. We can never have enough of them.

* * *

From all this, it follows that as we worship the success of the show business idols, we also watch with fascinated interest their missteps and failures. In some cases, these do them no harm, for their position seems to put them above the requirements of ordinary morality. Their extramarital affairs, their repeated divorces and remarriages, are the very food of our own fantasies. Oh, if we could only defy conventions as they do.

More malevolently, there is keen interest on our part in less happy violations of social mores. We follow tales and whispers of alcoholism and of drug addiction with a readiness to believe. We quickly accept accounts of professional jealousies and hatreds, of tantrums on the set, of decay of talents—almost as though, having given up any dreams we might have had of attaining our idols' place in the sun, we are only too delighted to see them descend to our own level, or below—thus punishing them for having dared to be famous and rich and happy.

The most violent reversal that a show-business personality can suffer is to be found guilty of a crime, especially the quintessential crime of murder! Most things can be forgiven someone who is famous enough and who is sufficiently adulated, but not murder. Murder brings not only the frisson of violent death to our senses, but the horrified thrill of knowing that a lofty career at once be ruined, that a god or goddess will fall from heaven.

In real life, it would not be kind of us to feel like this, but in fiction we can indulge our wicked impulse and yet do no one harm. Here in this anthology, "Show Business Is Murder" then, we bring you a score of tales that combine show business and murder, and we can enjoy ourselves freely.

What's more, we have tried to bring you a variety of aspects of show business. The characters in these stories are by no means leading men and women. There are also characters who are show-business has-beens, or who are screenwriters, or agents, or stuntmen, or vaudevillians, or (in the case of my own story) a piano player in a saloon.

Show business, after all, especially these days, takes in a very broad spectrum of activity, and it may well include even those industrious persons who sweep up after the elephants at the circus.

# 69

# Super

Once, some forty million years ago, there lived a little creature named "Eohippus." (Its proper name is "Hyracotherium," according to the paleontologists, but I like "Eohippus" better.)

Eohippus was a delicate little creature, with a stubby little horse-face, that browsed on leaves and scampered away as fast as it could when a meat-eater showed up. It was about the size of a fox, and it weighed perhaps as much as nine kilograms. It was a hoofed animal, with four little hooves on each foreleg and three on each hind leg.

If we suppose that the Eohippus could think after our fashion, we might ask him: "What do you suppose a Supereohippus would be like?"

It seems to me the little creature would answer, "Well, first it would have to be large and strong, so that it wouldn't be afraid of those rotten predators that are always chasing me. It ought to weigh a lot, maybe as much as thirteen hundred kilograms, so that it would be 140 times as big as I am.

"Then it would have to be tall, so that it could look far out over the plains and see predators if they come. It shouldn't be a quarter of a meter high at the shoulders, as I am, but perhaps as much as two and a quarter meters high—nine times my height—and it should have a long neck that would raise its eyes higher still.

"Oh, yes, and it should be fast so that if predators happened by, predators that were large enough to be dangerous or that hunted in packs, it could run like the wind and get away. It should run maybe as fast as fifty-five kilometers per hour, at least for short distances—just long enough to get away. Now *that* would be a Supereohippus."

Well, an Eohippus capable of saying all that would display a really active imagination, but even more so, real prescience, for such a Supereohippus actually lives today. Some of them are as massive, some as tall, and some as fast as Eohippus's dreams would have them. Supereohippus is what we, today, call a horse, and the word Eohippus itself comes from Greek words that mean "dawn-horse." That dawn-horse slowly evolved over the millions of years into today's supercreature.

Another example:

About three and a half million years ago, there were animals living in Africa that looked a little bit like small apes, but they were about 1.2 meters tall and delicately built. They walked erect, as we do. The animal is called "Australopithecus afarensis" by paleontologists. The first word is from the Greek and means "southern ape," because the first specimens were found in South Africa. The second word is from the name of a region in eastern Ethiopia where a particularly early skeletal specimen was discoverd in 1974. Creatures belonging to this and to related species are lumped together as "Australopithecines."

The Australopithecines are examples of "hominids" (from a Greek word for "man"), since their ability to walk erect puts them closer to modern human beings than to apes.

If an Australopithecine could reason as we do, we might ask it to describe its notion of a Superaustralopithecine. It might answer thus:

"Well, first of all, I'd want it bigger and stronger than I am, so it could defend itself better and be a better hunter than I am. I suppose that if it were, say, 1.6 meters tall, instead of 1.2 as I am, and if it weighed about seventy kilograms, instead of thirty as I do, it would be tall enough and strong enough to be a Superaustralopithecine.

"It would be no use making it even bigger and stronger, because that's not the chief advantage it ought to have. Some extra height, mass, and muscle is all right, but what it would mainly need is brains. My brain is pretty big for my size. It's bigger than that of the apes, which are stronger than I am. I'm doing better than they are because I'm smart enough to use tree-branches and thigh-bones as tools and weapons. The apes just use their nails and teeth.

"Now, then, it seems to me that if you really want a Superaustralopithecine, what you mainly need is a larger brain, one that weighs, say, 1.5 kilograms, or four times the size of mine. I realize that would mean

Superaustralopithecine's skull would have to be huge and swollen and he or she would look ugly, but there's no way out, if you want someone who is super. A big-brained Superaustralopithecine . . . who can possibly tell what one would be capable of?"

The Australopithecines were extinct by about a million years ago, but by that time there were hominids who were larger and bigger-brained. Eventually, there were still other hominids who were even larger and still bigger-brained. Indeed, about fifty thousand years ago a creature that was precisely the Superaustralopithecine described by our mythical australopithecine thinker made its appearance in Europe. We call it *Homo sapiens,* or "modern man." *We* are Superaustralopithecines.

Well, then, now it's our turn. We are actually the first species on Earth capable of deducing the evolutionary process that produced us, and the first capable of imagining a further evolutionary advance over ourselves. What would Superman be like? (Let me assure you that I am using the word "Superman" here, and in the title of this anthology, as a generic term to include *all* superhuman beings—women and children as well as men.)

The classic "Superman," the person who first appeared in *Action Comics* nearly half a century ago, is a good example of what I mean. He is not grossly different from us; indeed, he can masquerade as a "mild-mannered reporter" named Clark Kent. However, he is incredibly strong, has X ray vision, can fly through the air, and displays many other remarkable abilities. His great deficiency is that he is not particularly brighter than we are. In fact, he resembles the Siegfried of Wagner's operas who, according to Anna Russell, is "very handsome, very strong, very brave, and very, *very* stupid."

That is not exactly what we want. We want Superman to have a better brain than we have, to think and reason more efficiently; we would want him to have faster reflexes than we have, sharper senses, perhaps some senses we lack altogether. Then, if on top of that he is also stronger—fine, but that doesn't come first.

However, why puzzle it out? In *Supermen,* which I helped edit, we have an anthology of a number of excellent stories in which human beings are described who, in one way or another (and no two stories exactly duplicate the particular way), are markedly superior to ourselves. Sometimes the superiority rests upon the cooperation of a human being with another species, sometimes as the result of education and training, sometimes as a result of mutation.

It doesn't matter. The result is, in each case, the vision is of something like ourselves, but more remarkable. You can decide for yourself whether each particular vision resembles what *you* would consider super—or even endurable.

It may be, for all I know, that there are as many different kinds of Supermen as there are dreamers who wish they were something better than they are. My own idea of a Superman, for instance, would be someone who could write as well as Shakespeare, and who didn't need sleep, so that he could write *a lot* and—oh, yes, who would be infinitely attractive to women (at such time as he wasn't writing, and when his wife was willing to be cooperative—which isn't likely).

# 70

# Larger Than Life

How we love things that are larger than life! We insist upon them!

I don't know why that should be, but, after all, each of us started life as infants, and that may contribute part of the explanation. Our baby universe was filled with Mother and Father, who were far, far stronger than we were and possessed powers of such vague magnitude that they were effectively infinite. It was to them we turned for satisfaction and protection, and it may be that our first great disillusionment in life was the realization that they were not larger than life after all. As we get larger, and stronger, and perhaps wiser, we cannot avoid reaching that conclusion, however reluctantly.

No matter how we close our eyes to the truth, our parents *will* grow old and feeble and come to depend on us, and then, in ultimate betrayal, they will die, even if they avoid accidents, and no matter how well we care for them.

We cannot replace them. No one else is quite like the mother and father childhood knew. But we cannot make do without them, either, so we fall back on our imagination.

I wonder to what extent the myriads of gods that humanity has invented owe their existence to the necessity of possessing fathers and mothers who are *forever* larger than life and who will *never* betray their role by falling into impotence and death, as human fathers and mothers do.

Deities grow too perfect, though, and distance themselves from humanity to the point where they become etherealized into insubstantiality. For literary purposes, demigods are more satisfactory. They are larger than life, yes, but not so much so that they cannot suffer pain and occasional defeat. They are larger than life but *remain one with us.* (At that, even gods were most popular when they were human enough to suffer death, at least temporarily—Baldur, Tammuz, Adonis, and so on. To be sure, they symbolized the winter-death of vegetation, but the touch of humanity implicit in their death endeared them to their worshipers, and their eventual resurrection gave hope that death might after all be defeated and that separation by death might not be permanent.)

The first epic we know of is about five thousand years old and is of Sumerian origin. It is the tale of Gilgamesh, King of Uruk, who is mightier and more daring than a human can be. He is someone with whom we can vicariously share greater-than-human deeds, run greater-than-human risks, and suffer greater-than-human torment. Each culture creates its superhero. The Greeks had Heracles, the Hebrews had Samson, the Persians had Rustem, the Irish had Cuchulain. Each carved his way through a hostile world by means of his mighty thews, which more than made up for his usually less-than-subtle mind.

Oh, occasionally a superhero vanquished because of the subtlety of his mind: witness Odysseus. Generally, though, if wisdom that was larger-than-life was required, it made itself manifest through a knowledge of magic, as was the case with the Welsh Merlin or the Finnish Vainamoinen.

And no matter how accomplished the tales might be of wizards and tricksters, nothing moved readers like the men with mighty muscles. It was the immediacy of the hewing sword that counted. Perhaps the ordinary reader could more easily identify with strong muscle than a subtle mind—there was a chance one could develop something like the first, but one had to give up on the second.

Throughout history, the larger-than-life muscular heroes continued to invade the literary fields. In the Middle Ages, we had King Arthur and his knights, with the ever-victorious Lancelot as the acme of a chivalry that never existed in real life. And there was King Charlemagne and his paladins, with Roland as its exemplar. (The subtle trickster who was shrewder-than-life also existed, as in Reynard the Fox, Till Eulenspiegel, and so on; but again, they never quite had the same appeal.)

Then came the day when gunpowder ruled the world and muscles and armor were no longer of use; when a cowardly, weak-muscled, low-born wretch, by taking aim could clang Sir Lancelot to earth with a neat little hole drilled in his breast-plate.

Alas for heroic fantasy. Had it not died?

Modern literature makes up for that death by giving us other varieties of magnification. We still retain the shrewder-than-life protagonists of mystery novels—the Sherlock Holmeses and the Hercule Poirots. We have the more-beautiful-than-life heroines and heroes of romances, and the more-dreadful-than-life menaces of the Gothics and the horror tales, and so on indefinitely. We even make use of gunpowder for the purpose of inventing the faster-than-life Western heroes, who puncture the villains even after allowing them to begin to draw first.

But nothing substitutes for the more direct form of violence. In every form of literature, we end up with fistfights—a form of combat that is met with surprisingly un-often in real life. Detectives fight, Western heroes flail away, romantic lovers indulge in pugilistic displays. This is especially true in movies and television, where they leave no bruises and muss no hair, though the sound of bone on bone is deafening.

And even that is not enough. We still want the old pre-gunpowder days, when mighty biceps were needed to raise mighty swords and when the hero had to find a way of defeating sorcery with nothing but brute muscle at his disposal.

So why not write the story? It doesn't have to be in the present world, does it? It can be in the past. In fact, it doesn't even have to be in the real past, where it will be bound (however faintly) by known historical facts. Create a world of medieval civilization immersed in a sea of barbarism, and bring forth Heracles anew. You have "sword and sorcery." You have "heroic fantasy." You have all the dreams back.

Or you can add a little spice and variety to the mix. You can add a touch of advanced science, a soupçon of deliberate anachronism, a sprinkle of wry humor. The variations are endless, and the opportunities for the free-wheeling imagination are infinite.

Here we have thirteen fantasy novellas long enough to give their authors scope to invent intricate societies in which to display larger-than-life heroes and heroines facing larger-than-life evils and cruelties, suffering larger-than-life defeats, and winning larger-than-life triumphs.

You, caught amid it all, will live it all with them, and enjoy larger-than-life pleasures.

Remember, it's the oldest form of literature in the world, as old as *Gilgamesh;* and older, too, for heroes and heroics were probably celebrated by bards about campfires through thousands of years before writing was invented and probably ever since *Homo sapiens* has existed.

After all, from the very beginning of human history parents died—and had to be replaced.

# 71

# Science Fiction Mysteries

Mysteries and science fiction are two branches of literature that did not become possible, in their modern sense, until human society had developed to some appropriate pitch.

Thus, you couldn't very well write mysteries until society had developed organized police forces to combat crime. Until then, the occasional crime-detection story was enjoyed, but these were few and out of the ordinary.

Again, you couldn't very well write science fiction until society had developed science and technology to the point where change was fast enough to be visible and the notion of an advanced future began to make sense. Until then, the occasional fanciful story was enjoyed, but these were few and out of the ordinary.

It follows that the modern mystery story and the modern science-fiction story both date back to the early nineteenth century. As it happens, each received its most important early impulse from the writings of one man: Edgar Allan Poe.

In other respects, though, the two genres are different, even antithetical.

The mystery story represents the triumph of order. Crime (particularly murder) upsets the social fabric. It introduces an element of anarchy. Someone has deliberately deviated from the accepted code of social behavior to seek some sort of satisfaction through unsanctioned methods.

Presumably such an action, however small in itself, threatens us all, since if it is uncorrected, unpunished, and unavenged, it will open the door for further such actions and end by destroying society.

The protagonist of the story then must discover the nature of the anarchic action, the circumstances, the individual or individuals who attempted to pervert society. He or she must reveal the culprit and bring about suitable punishment. Order is restored and the social fabric is saved, so that the story ends exactly where it began.

The science-fiction story, however, represents the triumph of disorder. A science-fiction story must be set against a society significantly different from our own—usually, but not necessarily, because of some change in the level of science and technology—or it is not a science-fiction story.

This means that, to begin with, the science-fiction story destroys our own comfortable society. The science-fiction story does not deal with the restoration of order but with change and, ideally, with continuing change. In the science-fiction story, we leave our society and never return to it.

In fact, if we were to return to our society, if order were to be restored, the science-fiction story would be a flat failure. Imagine a science-fiction story in which the nuclear bomb is invented, in which its dangers are recognized, and in which the hero succeeds in suppressing the knowledge so that everything continues as before. That is not the way things work. When Pandora's box is opened, whether for good or evil, the world changes. The science-fiction writer may seek solutions and even find them, but there is one forbidden solution: forcing everything back into the box.

Can there then be a fusion of these two types of story—that of quintessential order and that of quintessential disorder? To be sure, mysteries can be written in which science plays a factor. Arthur Conan Doyle wrote of Sherlock Holmes, a scientific detective who was always peering through his magnifying glass at bits of tobacco ash. R. Austin Freeman's Dr. Thorndyke is an even better example of the scientific detective, and scientific minutiae often play a role in the mystery stories of the classic type.

In the same way, science-fiction stories often have a mystery motif, as in Lewis Padgett's "Private Eye" or Alfred Bester's "The Demolished Man." Usually, though, the science fiction is totally dominant and the mystery would not stand on its own.

John W. Campbell, Jr., science fiction's greatest editor, maintained that a perfect fusion was impossible. In 1953, to prove him wrong, I wrote "The Caves of Steel," a science-fiction mystery in which each element is equally strong, and in which each supports the other. I then wrote "The Naked Sun" and "The Robots of Dawn" as sequels, just to show that the first was no accident.

I am not the only one who does it, and here is Andrew Weiner's *Station Gehenna* to demonstrate that.

# 72

# The Science Writer

You might suppose that a science writer—especially one who is a correspondent for a daily newspaper—is in much the same position as a sports writer, the person who runs the society column (if there is such a thing anymore), or the drama reviewer. The science writer, it might seem, is just another one of those people who interprets a specialized field for the interested but inexpert onlooker.

Yet there must be more to it than that, simply because there's more than that to science itself.

It's quite possible that there are far more people who wonder whether the new play that's opened is any good or not than ever wonder about any phase of science, and far, far more than either who would like to know who's going to win tomorrow's baseball game—but importance is not the sort of thing that can safely be left to majority vote.

A drama misjudged, or a baseball victory unforeseen, may be the cause of much anguish, and even some monetary loss, but if science goes wrong these days, the result might just possibly be a catastrophe before which the most expensive of wasted tickets and lost wagers must pale.

Again, the world faces an appalling list of life-and-death crises these days, life-and-death for our entire industrial civilization and, not so incidentally, for some billions of those who make it up and who (both themselves and in the person of their recent ancestors) have flourished and multiplied under it for over two centuries.

The sad litany is so well-known that it is scarcely worth going through it: dwindling resources, including, most particularly, oil; gathering pollution; increasing weight of overpopulation; vanishing topsoil and expanding deserts; multiplying nuclear weaponry; deepening pin-pricks of terrorism and social alienation; intensifying popularity of various cults of the irrational.

What solutions to these can we imagine? Pehaps none, but if there are to be any at all, they will arrive through advances in science and technology, which can (or at least *might*) supply us with new energy sources, better means for recycling and conservation, truer understanding of reproductive physiology and psychology, more efficient means for detoxifying the environment, and so on.

*If* science and technology are allowed to advance, that is.

It is by no means certain that they will be. A great many of Earth's problems arise from the previous success of science and technology, because that very success made unwise excesses so easy and so attractive. Technology has too often been driven hard by the short-term goals so clearly in view, to the (much less visible) long-term detriment of humanity. And the very people who cheered on the folly as long as they profited now turn savagely on science when the consequences come to be weighed and measured. What else is there to blame, the mass-greed and mass-ignorance of millions?

It is easy, in any case, to feel disillusioned about science—which is rational and cautious by its very nature, and which is only able to tell us that it seems likely, on the evidence, that two and two is four. How much more fascinating and comforting are the fringe beliefs that assure us, in the most confident and heartening matter possible, that two and two is certainly equal to six and a half.

Under these circumstances, it might appear that humanity doesn't have a hope.

If, as it seems to me, the only reasonable chance we have of pulling out of the hole into which we have dug ourselves is to encourage scientific and technological advance and use that advance with measured caution and judicious wisdom. . . .

And if the record of the last couple of centuries is such that we seem much more likely to use the advance, as always, for immediate short-term comfort and profit, without consideration for the long-term consequences . . . and if there is, in any case, growing disillusionment with this miserable record, and, in our compounded folly, we turn away not from our own unwisdom but from the principle of science and rationality . . . then, surely, we face ruin.

Except that it goes against the grain of human beings to give up and accept defeat. Perhaps, if those of us who can, patiently explain the principles of science, describe the latest advances, point out the fallacies in the silly pap that is fed the public by cleverly irrational knaves (or sincere ignoramuses), then enough people might be swung to the side of rationality and wisdom to save us yet.

In 1920 H. G. Wells said, "Human history becomes more and more a race between education and catastrophe," and in the sixty-nine years that have passed since, education seems to be losing the race. Ironically, as our growing scientific and technological capacity make ignorance more and more the likely agent of catastrophe, it is scientific education (of all things) that falls farthest behind.

The more reason to redouble our efforts!

Science writers, especially the science correspondents of great newspapers, face the most difficult and the most important part of that task. They face the public, the great mass of humanity, far more people than any academic in the classroom does or can.

Science writers have an audience that can, all too easily, skip their essay and turn to the advice-to-the-lovelorn column. And if they can manage to seize some of that audience, they must often tell them what they don't want to hear—that what they would like to believe is mere nonsense leading them astray.

Yet they *must* do all this, and if they can succeed even a little bit, we are that much farther from the edge of the cliff.

It is consequently heartening, at this point, to be able to present a collection of Adrian Berry's essays from *The Daily Telegraph*. Adrian knows what he is saying, he writes well and interestingly, he makes matters plain, and he is fearless in assaulting the forces of irrationality and superstition.

And he is on the side of the angels—for reason, for the advancing front of knowledge, for survival and growth, for the magnificent heights of the future.

# 73

# The Scribbling Scientists

One of the great scientists in the days when modern science was beginning was the German astronomer and physicist Johann Kepler (1571–1630). His fascination with the heavenly bodies was such that he actually sat down and wrote a romance about someone who had flown to the Moon. He was cautious enough to keep it hidden during his lifetime, though, and it was not published until 1634.

Interplanetary travel was not totally original with Kepler. The first fictional trip to the Moon that still survives today was written by Lucian of Samosata in Roman times, fifteen centuries before Kepler. Kepler, however, was the first scientist to do so. In fact, he was the first scientist who, as far as we know, wrote a story of any kind that was recognizable science fiction.

This created problems for Kepler. Earlier authors who wrote of flights to the Moon were romancers who did not trouble their heads with such matters as the distance of the Moon, which was known to be enormous since the second century B.C. Lucian had his travellers reach their goals by being lifted there by a water-spout.

Kepler was too knowledgeable to use such a device, but failing to think of a rational method, he had his hero carried to the Moon by spirits. Where earlier writers treated the Moon as just another country, not much different from such little-known regions as Africa or India, Kepler knew that the Moon was different. He knew, for example, that it had a day and a night that were each two weeks long. He therefore imagined the Moon to be populated by strange plants and animals that grew madly during the day and died at nightfall.

Kepler did not really set a precedent. Slowly, science fiction grew and developed, but those writers who nurtured it, though usually quite knowledgeable of the science of the day (Edgar Allan Poe, Jules Verne, and H. G. Wells were all scientifically literate), were not *professional* scientists.

Nevertheless, occasional scientists did try their hands at it, even though they usually didn't set the literary world on fire with their efforts. (This is not surprisng. The craft of writing often requires a long and painstaking period of apprenticeship, and professional scientists have arduous work of their own and usually lack the time to develop literary polish.)

The French astronomer Camille Flammarion (1842–1925), who was an ardent believer in life on other worlds—especially Mars—tried his hand at science fiction. The Russian physicist Konstantin Tsiolkovsky (1857–1935), who was the first to deal with rocketry and space flight with mathematical rigor, also wrote a science-fiction novel to illustrate his ideas for those who were not up to dealing with his equations.

Nowadays, however, it has become quite common for scientists to try their hands at science fiction. Why not? A surprising number of scientists found themselves fascinated by science fiction in their younger days, and some were even lured into science in this way. Why not, then, return the compliment, so to speak, and try to fascinate others in the same imaginative way that they themselves had been caught?

It's not easy to do this, however, for to be a great scientist who writes great science fiction, one must have achieved an extraordinary balance, rather like a cone standing upright on its tip without falling over. To tell you the truth, I know some great scientists who write science fiction, and some great science-fiction writers who are scientists, but, despite the title of this anthology, I don't know any *great* scientists who are *great* science-fiction writers.

The trouble is, greatness in any direction doesn't usually catch a person by surprise. If, early on, you realize you are likely to become a great science-fiction writer, you find yourself fascinated with your writing and, even though you may be on your road to science, you are all too likely to abandon that road. Again, if, early on, you realize you have the capacity to be a great scientist, then even if you have the urge to write science fiction, you never manage to find enough time to do much of it.

As a matter of fact, I myself came closer to the balance than most people. From my middle-teens, I was determined to be a science-fiction writer, and I was also determined to be a scientist. I moved ahead in both directions with a kind of ferocious energy. I wrote more and more science-fiction stories (and some are now recognized classics), while at the same time I obtained my Ph.D. in chemistry from Columbia and then joined the department of biochemistry at Boston University School of Medicine. By 1958, I was an associate professor and had written textbooks—and was at the same time in the first-rank of science-fiction writers. But by then, I had to choose. There was no time to do both properly. I chose writing. I still have my academic title, but I have been a full-time writer since 1958.

Nevertheless, it is important, I think, to emphasize that there *are* people who try to do both, for too many scientists and too many science-fiction readers think that there is some kind of peculiar dividing line between science and science fiction—that scientists must, of necessity, scorn science fiction, and that science-fiction writers must, of equal necessity, be ignorant of science.

Not so! *Not so!*

So here we have a sizable collection of science-fiction stories of excellent quality that have been written by people who are either practicing scientists or who obtained the education to become such but drifted away because of their greater interest in writing.

Wherever possible, we chose a story in which the writer dealt with his own field of expertise. Thus, my story "The Winnowing" is one of the few I have written that deals with biochemistry. We do not insist on the hard sciences only. Mario Pei (whom I knew when he was alive) has a story that deals with linguistics in a thoroughly delightful and almost de Campian way.

So I wish you all joy in in reading *Science Fiction by Scientists.*

# 74
# Neanderthal Man

In Western Germany, in the middle course of the Rhine River is the city of Dusseldorf. Directly to its east, along the banks of the small Dussel River is the Neander Valley. Naturally, the people of the region gave it the German version of the name. The German word for valley is *tal,* or in more archaic spelling *thal* (though it is pronounced the same either way). The region east of Dusseldorf is, therefore, the *Neandertal,* or *Neanderthal.* The Germans pronounce it "nay-on′der-tol′ " and we pronounced it "nee-an′der-thawl."

In the Neanderthal in 1857, workmen were clearing out a limestone cave, and they came across some bones. This is not an uncommon thing, and most of the bones were thrown away. Fortunately, some long bones were preserved together (very fortunately) with the skull.

The bones were clearly human—but there were significant differences. The skull had pronounced bony ridges over the eyes, which ordinary human skulls do not have. It also had unusually prominent teeth and a receding chin.

The discovery was made two years before Charles Darwin published his great book on evolution, but there were evolutionary ideas in the air. Some people were bound to wonder whether this "Neanderthal man" was perhaps a primitive ancestor of modern man.

Since most Europeans at the time (including scientists) were convinced that the early chapters of the Bible were literally true, this evolutionary notion met with strong resistance. Many people insisted that the bones were quite modern, and one suggestion was that they were only a generation old and were the remains of a Cossack soldier who had died during the Russian march into Western Europe in pursuit of Napoleon.

In 1863, however, Thomas Henry Huxley, a great champion of Darwinian notions, studied the bones and came out strongly in favor of their being the remains of a primitive species of man. In 1864, another British scientist, William King, named this primitive species *Homo neanderthalensis.*

If the Neanderthal find had been the only one of its kind, the dispute might have raged to this day without settlement. However, other sets of Neanderthal skeletons were found by the dozen, and there is no doubt now that they are primitive precursors of modern man.

In 1911 a nearly complete skeleton of Neanderthal man was studied by a French scientist, Pierre Boule. It was his description that gave rise to the popular picture of Neanderthal man as a short, brutish, shambling, grotesque creature of ape-like appearance.

As it happens, though, Boule was working with the badly arithritic and deformed skeleton of an old man. The study of other skeletons of younger individuals in better health make it seem that Neanderthal man was a lot less subhuman than that. Yes, there are the heavy brow ridges, the large teeth, the receding chin, the retreating forehead, but on the whole Neanderthal man stood bolt upright, walked exactly as we do, and was not markedly different from us from the neck down.

What's more, the Neanderthal brain is as large as ours and even, perhaps, a little larger, though it is differently proportioned. The Neanderthal brain is smaller in front (hence the retreating forehead) and larger behind. Since the front part of the brain is associated with the more rarefied regions of abstract thought, we might suppose that the Neanderthals were less intelligent than we—but there is no real evidence of that.

If Neanderthal man is not too different from us, might individual Neanderthals have intermarried with individual "modern" human beings? They not only could, but they apparently did, for skeletons showing features intermediate between Neanderthal and modern men have been found. It is now considered that Neanderthal men belonged to the same species we do, so that their scientific name is now *Homo sapiens neanderthalensis,* while we are *homo sapiens sapiens.*

Neanderthal man may have lived on this Earth as long as 250,000 years ago and were the dominant form of human life for perhaps 80 percent of the time since. Modern man may have become prominent only 50,000 years ago, and it may not have been until 30,000 years ago that the last human being of pronounced Neanderthal features died.

But perhaps not quite. The popular feeling is that Neanderthals were wiped out by the "superior" tribes of modern men who multiplied and grew powerful, but it may also be that the extreme Neanderthal features were blurred and erased through interbreeding. Perhaps Neanderthal genes flourish within the present-day human population, especially in Europe, where Neanderthals seem to have been most numerous. My dear wife, Janet, for instance, is convinced that in addition to her Viking ancestry, she also possesses numerous Neanderthal genes, which she considers as representing a sign of noble origin. (She has no proof of this, but the thought pleases her, so I wouldn't dream of arguing.)

And after all, why not be proud of Neanderthal descent? The Neanderthals managed to survive the Ice Age; they were Stone Age men (as were modern men for a long time) who used fire, had a variety of useful tools, probably led a complex social life, and even showed spiritual yearnings, as we judge from the fact that they buried their dead with flowers and with utensils that might presumably be of use in an afterlife.

To this day, the view of Neanderthals—as expressed in science fiction, for instance—varies between showing them as definitely inferior, as in Lester del Rey's "The Day is Done," or as almost indistinguishable from us, mentally, as in my own story, "The Ugly Little Boy."

Whether mentally inferior or not, the attitude toward Neanderthals tends to be rather sympathetic these days, as, for instance, in Jean Auel's best-selling *Clan of the Cave Bear.*

In any case, here in *Neanderthals* you will find a number of stories considering various aspects of what might have been the Neanderthal experience. They differ among themselves widely, of course, but I am certain you will find them all pleasurable.

# 75

# The Nonhuman Brains

Through all of history, human beings have been able to exchange ideas by way of the spoken word only with other human beings. Intelligence of the human sort has existed only in ourselves, and that seems to have made us lonely.

At least very many of the ancient myths and legends of humanity deal with speaking beings who are not human.

In the Bible, for instance, there are two speaking animals. In chapter 3 of Genesis, there is the serpent who speaks to Eve and persuades her to eat the fruit of the tree of knowledge of good and evil. In chapter 22 of the Book of Numbers, the donkey on which the wizard Balaam rode was given the power of speech in order that it could protest having been beaten for trying to avoid an angel that barred its path, an angel Balaam himself did not see.

In Homer's *Iliad,* which ancient Greeks revered as modern Westerners revere the Bible, the Greek warrior Achilles rode horses that were divine and immortal. As he prepared for the final battle in the epic, one of those horses spoke and warned Achilles that he would survive the battle but would die soon after.

Animals that think like human beings and exchange ideas in the human fashion are to be found in all sorts of folktales, from the stories in Grimm to those of the *Uncle Remus* tales. Even modern storytellers think up such stories, as Hugh Lofting did in his *Dr. Dolittle* books and Richard Adams did in *Watership Down.*

Not only that, but all sorts of fanciful creatures of human or even superhuman intelligence have been invented. Fairies and elves were thought to inhabit the Earth with us. Demons and nymphs and monsters of many sorts were thought to surround us.

Nowadays, though, we are skeptical and aware that there is no evidence of any nonhuman intelligence about us. Animals do not truly speak. Even parrots and myna birds, which can mimic human sounds, do not understand what they say. Chimpanzees and gorillas may be able to communicate simple ideas in sign language, but they can't go very far in this direction. Dolphins may have a language of their own, but we seem to have no way of understanding it.

And as for spirits and monsters and demons and fairies—nonsense!

So human beings—their brains, the ideas those brains invent, and the speech that communicates those ideas—remain alone on this planet.

It may be that there are other intelligences equal (or even superior) to our own, on other planets circling other stars—many astronomers think there must be—but we have no evidence for that. If extraterrestrial intelligences exist, we have not heard from them or detected them. (People may talk about UFOs and about "ancient astronauts," but that, too, seems to be just nonsense.)

Well, then, are we condemned to intellectual loneliness forever?

Not at all, for human beings have set about creating thinking machines; they are designing nonhuman brains; they are building devices that represent "artificial intelligence."

The idea of human beings actually constructing something that can think is an old one. There is the medieval legend of the Golem and the modern tale of the monster built by Dr. Frankenstein, to say nothing of later science-fiction tales of all sorts. These, however, were just exercises of the imagination.

The first person who thought of a legitimate way of constructing a machine that would do some of the things that till then only thinking human beings could do was an English mathematician named Charles Babbage. About a century and a half ago, he tried to build a machine that would be capable of solving mathematical problems. Unfortunately, all he had to work with were mechanical devices—little wheels, gears, levers, and so on. He tried for many years, but he could never get his machine to work as he wished it to. Eventually, he ran out of money and had to abandon the project.

By the time of World War II, however, scientists had electronic devices to play with, and these could do the job much more quickly and delicately than mechanical devices could. In 1946 the first electronic computer, ENIAC, was built, and in the forty-three years since rapid advances have made such computers capable of conducting ever more complicated operations while growing smaller and smaller.

In the middle 1970s, switching devices were made of practically microscopic size (the "microchip"), and it was possible to have very complex computers so small you could slip one into your pocket. They are so cheap that almost everyone can have one.

What's more, everything is becoming computerized. I am writing this article on a computerized "word-processor." Automobiles are being built by computerized "industrial robots." Children are playing with computerized "video games," which in just a couple of years have become a multibillion dollar industry. Who knows where it will all lead?

What if computers grow to be as intelligent, in their own way, as human beings? Or even more intelligent? They will be like the nonhuman intelligences we imagined in our legends. In that case, will they work for us or against us?

Will advanced computers try to save us, as the donkey tried to save Balaam? Or will they try to harm us, as the serpent tried to harm Eve? Will the computers be helpful elves or spiteful goblins? Good fairies or wicked enchanters? Angels or demons?

We don't know, really, but we can imagine, and it is science-fiction writers who, in such directions, do most of the imagining these days. In *Computer Crimes and Capers,* therefore, we present science-fiction stories that deal with what might happen if computers grow much more capable than they are already.

The general fear seems to be that computers might get out of human control. Well, think about it. Might they really? And if they might, how does one prevent it? Think about it right now—then read the stories and think again.

# 76

# Computer Envy

As we age, we tend to realize that the various parts of our body are wearing out. This is sad, but inevitable. If the reader happens to be young and finds this difficult to believe, all I can say is "Wait!"

We shouldn't really complain about it. Inanimate objects and human artifacts that have no moving parts (statues, for instance) may exist, reasonably unchanged, far, far longer than we do—at the price, however, of doing nothing.

But anything with moving parts *ages*—and usually far faster than we do.

Among living things, no mammal lives as long as human beings do, on the average. And the only animals that do better than we (turtles, for instance) are cold-blooded and live quite slowly. Plants live longer still, but they live even more passively.

Inanimate objects with moving parts do poorly, too. A watch, or a washing machine, or an automobile that is as old as I am and has not been repaired, is not likely to be working at all.

However, the point is that they *can* be repaired. Little by little, you can replace this part of an automobile, then that part—the tires, the engine, the headlights—until no part is precisely the part that existed when the automobile was bought and yet there will be a continuity about it.

Why can't the body do it?

The body *does* do it, of course. Cuts and lacerations heal, broken bones knit, and so on. The capacity for such things decreases with age, however, and eventually the body wears out in ways that won't spontaneously heal, so that even if you avoid infection or accident, you die of some form of degeneration or other.

Technology comes to one's aid. Teeth decay—the only part of the living body to do so—and that decay is irreversible. In the old days, that meant the eventual loss of teeth, all of them. Now, metal amalgams fill the cavities left by rot. Capping, bonding, and root canal work make it more likely that teeth will not be lost.

I, for one, have glass lenses in front of my eyes to supplement the natural lenses within them. In fact, when my coronary arteries grew dangerously plugged a few years ago, surgeons ran replacement arteries and veins around the plugged portions (using parts of my own arteries and veins for the purpose, to be sure) in order to see to it that my heart continued to receive an adequate blood supply.

Obviously, it would be nice if we could do still more. There is the Jarvik heart, which can pump away for a while during the wait for an organic replacement to show up. Wouldn't it be nice, though, if we had a more permanent mechanical heart with a self-contained energy supply that could last for centuries?

Ditto, ditto, for liver replacements, eye replacements, and so on.

In fact, in my story *Bicentennial Man* I had in the foreground a robot that came to be more of a man until he developed the final human ability of degenerative dying. In the background (I didn't make much fuss over it) were human beings who were learning to add more and more to themselves in the way of prosthetic devices in order to avoid, or at least, delay degenerative dying.

The totally unexpressed idea was the robots would become more like human beings and human beings more like robots until, finally, one couldn't tell the difference.

These ideas are not new, of course. They certainly antedate science fiction. In *Pinocchio* we have the case of a living marionette (as close to a robot as you can expect to get) who succeeds in becoming a boy through his loyalty and bravery. For the other way around there's *The Wizard of Oz,* where a woodman who is unbelievably clumsy accidentally mangles his various limbs with his axe and is rebuilt out of metal, so that we have the Tin Woodman, who is clearly a more permanent version of himself.

If we go back to the Greek myths, there is a trace of the mechanical replacement of parts. In one of the less pleasant myths, Tantalus, a king of Argos, was described as an intimate of the gods. Once, when he invited the gods to a banquet at his palace he determined to test their omniscience by killing his son, Pelops, and presenting his remains as food for the banquet.

The gods realized what he had done, of course, and restored Pelops to life and arranged to have Tantalus tortured in Tartarus. There he was forced to stand in water up to his neck, with delicious fruit waving before his face. But when he stooped to drink, the water level sank, and when he reached for the fruit, it swayed away (hence the word "to tantalize").

However, at the banquet the goddess Demeter, preoccupied by the loss of her daughter Persephone to the god of the underworld, was paying no attention and absent-mindedly ate a portion of Pelops's left shoulder. The gods therefore replaced the missing part of the body with an ivory prosthesis—which was rather neat, I think.

But would human beings be content to have their body supplemented or even replaced by mechanical analogs? Personally, I think they would. They do not reject such things now. They do not refuse, on principle, to have their teeth filled, or to wear glasses for improving vision, or to accept replacement hip joints, or pacemakers, or artificial legs, and so on.

In fact, I suspect that if these devices were markedly more efficient or longer-lasting than the "real thing," people would line up to have them. Gardner Dozois, harking back to old Sigmund Freud, calls it "computer-envy" in his book, *Antibodies,* and I think that's just the right phrase for it.

# 77

# Dogs

The Bible hasn't a good word to say for dogs. Thus, Jesus is quoted as saying in the Sermon on the Mount: "Give not that which is holy unto the dogs" (Matt. 7:6), where dogs symbolized anyone who is totally unworthy of divine blessing.

This is perhaps understandable, for in the Middle East in biblical times (and perhaps now, too) the dog was a scavenger and carrion-eater, a source of fleas and disease, very useful in the absence of a sanitation department but not much good otherwise.

In this respect, however, the biblical point of view does not represent the attitude of our Western world. In Homer's *Odyssey,* Odysseus returns from a long war under the walls of Troy and long years of wandering since. He comes home after a lapse of twenty years to find his wife still faithful but besieged by arrogant suitors who are trying to succeed to the kingdom by marrying her.

The returning king arrives in the guise of a beggar, and of course no one recognizes him (except his old nurse, who sees a distinctive scar on his leg, which he received while hunting a boar when still a youth). However, Odysseus had a dog whom he had bred before leaving for Troy and who had been a famous hunter. He was left behind when Odysseus left, and now, twenty years older, he barely retains life. He is lying on a dunghill, covered with fleas, but as Odysseus approaches, the dog, Argus, wags his tail and tries to stand up, but can't—and he dies even as he carries through this feat of recognition.

Odysseus weeps at this and who knows how many millions have read the *Odyssey* and felt a suspicious moistness about their own eyes. I myself feel it now, having just reread the passage, even though I seriously doubt that a dog would have recognized its master after twenty years.

The loyalty and utter fidelity of a dog has come down as an article of faith to most people of our Western culture. In 1884 Senator George G. Vest gave a speech in the Senate (a speech that, as far as I know, is the only deed for which he is remembered) and spoke highly of dogs. He said, in part, "The one absolutely unselfish friend that man can have in this selfish world, the one that never deserts him, the one that never proves ungrateful or treacherous, is his dog. . . . He will kiss the hand that has no food to offer. . . . When all other friends desert, he remains."

Again, my sense of cynicism keeps me from ignoring the fact that there are such things as conditioned reflexes and that a dog may be following these rather than any high moral sense of loyalty. In any case, Vest's encomium struck a chord in millions, and it may have given rise to the common expression, "A man's best friend is his dog."

Mark Twain, ten years later, said much the same thing in his book *Pudd'nhead Wilson,* but he did so in the bitter, epigrammatic way of which he was a master: "If you pick up a starving dog and make him prosperous, he will not bite you. That is the principal difference between a dog and a man." I suspect, however, that Twain said that not because he loved dogs so much as because he loved men so little.

Whether or not the dog is man's best friend, he seems certainly to be man's oldest animal friend. The dog has been domesticated since 8000 B.C. at least, and he is the one animal that has followed human beings over all the world. The American Indians had very few domesticated animals, but they had the dog. Even the Australian aborigines had the half-wild dingo.

Why is that? Dogs are useful, for one thing, since they are hunting carnivores and can therefore hunt along with human beings. Dogs are also intelligent enough to see that if they actually help human beings with the hunting they will be rewarded with a share of the carcass. It has been a symbiotic relationship that has helped both. Dogs could locate the prey, pursue it, worry it, wear it out, and men with spears and arrows could do the actual killing.

Dogs might have descended from some variety of wolf or, perhaps, jackal—we can't be sure which—and the association must have arisen because some early dogs would sniff around a campfire to see if they could scavenge any leftover food.

Human beings may not have felt very fond of dogs at first, since when a dog runs off with a precious gobbet of meat, that does not win its way into the heart of a human being who had been about to eat it him- or herself. There must have been wild shouting and thrown rocks.

However, my own theory is that we have to consider the young of the species. Every once in a while a human child would come across a doggish child (a puppy, in other words), and they might have enjoyed each other. When it came time for the parents to drive the dog away (or, perhaps, to consider it as an article of diet), the child would object vociferously—and children often have their way in such matters.

As a dog grows older and more dangerous, however, the pressure to get rid of it or kill it grows stronger. Those dogs escape this fate which show themselves to be particularly friendly and amiable, or particularly adept at hunting. In other words, without knowing exactly what they were doing, human beings would discard some dogs and keep others and be breeding into their animals friendliness and hunting ability.

It helps in this respect that dogs are pack animals and have the instinct of following their leader—that is, whichever of their group can either beat the others physically or browbeat them psychologically. If a puppy is brought up by human beings, one human being or another becomes the leader as far as that puppy (and, later, dog) is concerned, and his fidelity is, again, a matter of instinct rather than some high moral virtue.

But what is the use of talking of instinct and conditioned reflexes? That means something only if we can retain some sort of cool and aloof attitude towards dogs, and that is a difficult thing to do.

Whatever the cause, the dog acts toward human beings *as if* it is filled with love and affection for them, and that appearance of love is taken as fact, and the love is usually returned in full.

Man has many domestic animals, kept for food or for work. Love has nothing to do with it. Chickens, geese, cattle, and sheep show no love. They are at best docile, and they do what is expected of them, yielding eggs or milk or wool.

Even the hard-working horse, perhaps the most beautiful of animals, which can inspire the affection of its owner, can show only a nuzzling affection of its own, one that is quiet and subdued.

The only animals that are kept as pets *only,* kept only out of affection even when they do no work and have no use, are dogs and cats, and of these, cats show no affection and less often inspire the deep love lavished on dogs.

But why do I say "no use"? To experience a steady, unwavering constantly expressed love—is that "no use"? To have a companionship that does not fail—is that "no use"? Psychologists have recently discovered that the possession of a pet contributes enormously to mental health and to emotional stability. They are probably the last to have made the discovery. Everyone else knows it.

Why not, then, combine two great loves: the love for dogs and the love for mysteries, and present the readers with mysteries that in one way or another involve dogs? In *Hound Dunnit,* you have a collection of such mysteries.

# 78

# Dragons!

Human beings have invented many an imaginary monster, but of them all the most fearsome and impressive is the dragon.

In our Western tradition, the dragon is a long, large, scaly monster, with wings and fiery breath, malignant in nature, and deadly in deed. It is the supreme act of a hero (whether divine, semidivine, or human) that he slay a dragon. The Greek god Apollo slew one in establishing his temple at Delphi. The Teutonic hero Siegfried slew one, and so did the Christian hero St. George.

Dragons were worth slaying for several reasons. For one thing, they tended to guard precious hoards. In the Greek myths, a dragon guarded the Golden Fleece and had to be killed by Jason, while another dragon guarded the golden apples of the Hesperides and had to be killed by Heracles. Fafnir, the dragon Siegfried slew, also guarded a fortune in gold.

Other dragons had to be killed because they had an unpleasant habit of dining on virgins (who were highly valued, then as now, for their scarcity, and who were thoroughly wasted when merely an item on a menu). Or they could be the embodiment of evil, for in the Bible the dragon is mentioned two or three times as a primal enemy of God.

Dragons, in a broad sense, could be any large and fearsome monster—one, for instance, that inhabits the sea rather than the air. The sea-monster that threatened the beautiful Andromeda and was slain by Perseus in the Greek myths is sometimes spoken of as a dragon. So is Leviathan, the great sea-monster mentioned in the Bible.

Some dragons were abnormally monstrous, such as the many-headed Hydra, slain by Heracles, or the six-headed Scylla whose gauntlet Odysseus had to run.

On the other hand, the Chinese had dragons that superficially resembled the flying dragons of the Western tradition. Chinese dragons, however, tended to be beneficent spirits.

Where did all these dragons come from? How did they arise in the human imagination?

One possibility we can dismiss at once. They were not versions of the dinosaurs. The most fearsome land animals of all time were the huge reptiles of the Mesozoic. Knights who routinely faced and destroyed dragons in the medieval romances might well have quailed if they had to face an angry tyrannosaur. There were even flying reptiles, the pterosaurs, the largest flyers that ever lived.

However, all these monster-lizards died out sixty-five million years ago, and human beings had absolutely no knowledge of them until the nineteenth century.

Can we be sure? Isn't it possible that some survived into early human times and formed the basis of the legend?

No. The chances of survival are so small that we might as well utterly ignore the possibility.

What, then, was the birth of the dragon?

Undoubtedly, it started as an actually existing animal and was improved upon. For one thing, there is the snake. The word "dragon" comes from the Greek *drakon,* which originally referred to a sharp-eyed earth spirit of some sort and was later applied to snakes.

Snakes have unwinking eyes, and that gives them the impression of being able to see much, of being sharp-eyed. They slither noiselessly underfoot and, if poisonous, are apt to strike without warning when frightened. This makes them seem evil and malevolent. Furthermore, if a poisonous snake strikes, the bite, if not fatal, will inflame and pain badly, and it is not far from that to the notion of a fiery breath.

And snakes can be improved on. Exaggerate the poisonous nature and you get the cockatrice or basilisk, which kill not only by means of poisoned fangs but by poisoned *breath* . . . or even by a poisoned *look.* An easier exaggeration is to make the snake monstrous in size (and some snakes are indeed ten meters long). It is difficult to imagine monster snakes slithering along the ground, but they can easily be made even more dangerous if allowed to slither through the air on wings.

(The wings are often pictured as bat-wings, since bats—those creatures of the night—are also viewed as ill-omened creatures. It is perhaps the bat origin that makes dragon wings so oddly small in most illustrations; but then, what do the myth-makers know of aerodynamics?)

And the sea-dragons? What of them? The real animal that serves as the model there is surely the crocodile, the deadly monster of the Nile. The description of Leviathan in the Book of Job is almost surely inspired by the crocodile.

The crocodile, like the snake, is a reptile, and both contribute to the scaly hide of the dragon. If the breath of the airborne dragon is pure snake, the short legs it is often pictured as having are pure crocodile. (And contributing to the many-headed sea-monsters like Hydra and Scylla is, undoubtedly, the many-tentacled octopus.)

In ancient and medieval times, it was thought that dragons (as well as other fanciful creatures) really existed, and it wasn't totally irrational to think so, since so much of the world was undiscovered. Who knew what could exist in the strange, misty lands beyond the horizon?

In modern times, however, we know well that dragons don't exist and never existed (unless you want to count the pterosaurs).

Nevertheless, they are still to be found in the imagination, and the dread dragon Smaug in Tolkien's *The Hobbit* is no less fearsome for being imaginary.

On the other hand, the gift of nonexistence is this: We can, if we wish, make our dragons bumbling, well-meaning creatures, or even entirely kindly. There is Walt Disney's *Pete's Dragon,* where the dragon is rather like an overgrown puppy-dog; and the hero of *The Reluctant Dragon,* which only wants to be left in peace; and the altogether kindly and ill-used protagonist of *Puff, the Magic Dragon* of the affecting ballad.

In the end, then, dragons have become the most lovable, as well as the most frightening, monsters of all, and those of us who write science fiction and fantasy have the broadest possible range of plots—and dragons—to deal with. The stories included in *Dragon Tales* will amply demonstrate the point.

# 79

# The New Beginning

Every year has its beginning—but when?

You might think that it doesn't matter, that it is quite possible to choose some day arbitrarily and say, "Let the year begin at this moment." You then count the days, and every time you count off 365 days (366 in leap year) you begin a new year the next day.

That sort of arbitrariness is certainly possible, but events in the sky guide people to natural beginnings of the year.

The noonday Sun moves first higher, then lower in the sky, and it repeats the process over and over without any significant deviation. When the noonday Sun is at its highest point, that is the "summer solstice" (at least that is the appropriate name in the northern hemisphere, where most of the human population lives). On our calendar, the summer solstice comes on June 21.

When the noonday Sun is at its lowest point, that is the "winter solstice" (December 21). When the noonday Sun is moving upward and has risen halfway from the lowest to the highest point, that is the "vernal equinox" (March 20). When it is moving downward and has fallen halfway from the highest to the lowest point, that is the "autumnal equinox" (September 23).

The least important of these four days is the summer solstice. To be sure, the days are longest then, the nights shortest, the fields green, the people comfortable and happy. "Midsummer Day" is celebrated with fun and games, but it is no time for beginning the year.

The vernal equinox, however, comes when the winter is finally over, when spring is clearly on the way. Soon the buds will appear on the trees and the ground will turn green and it will be time to sow the grain. That is clearly the beginning, and some calendars (that of the early Romans, for instance) began the year on or about the vernal equinox. Passover and Easter are spring festivals and are associated with the vernal equinox, too.

The autumnal equinox is also a useful mark. It begins the autumn; the grain is ripening to the harvest; the animals have produced and raised their young. It will soon be time to collect the food on which human beings will live comfortably through the cruel winter. That will mean a good life for another year, and that is surely worth celebrating. Thus, to this day, the Jews begin their year on or about the autumnal equinox. The French Revolutionaries, establishing a new "scientific" calendar in 1792, did the same.

However, it was the winter solstice that, oddly enough, has had the strongest claim to serve as a beginning, even though it marks the beginning of winter. Ahead lies three months of frosts, biting winds, and cruel storms, so what is there to celebrate?

Ah, but on the winter solstice, the Sun reaches its lowest point and will fall no lower. It is now beginning to rise again, so that no matter how bitter the winter, each day the noonday Sun is higher and the coming of warm weather, of another spring, of another planting, of another harvest is guaranteed.

It is not surprising, then, that in the Western world the two greatest and happiest holidays are what we call Christmas and New Year's Day, which come hard on the heels of the winter solstice, on December 25 and January 1 respectively. Nowadays, we associate Christmas with the birth of Jesus, but there is no biblical warrant for that.

(Before Christmas took over, there were earlier celebrations of the solstice, having nothing to do with Christianity. Lovecraft's "The Festival" makes this point, though the story is unhistorical and quite overblown, after the typical Lovecraftian fashion.)

The Christmas-New Year week represents the "holiday season," a period when the world is immersed in good will and cheer. The cry on all sides is "Merry Christmas" and "Happy New Year." Yet even so, it is well-known that the holiday season is hard on many people. The cry of "home for Christmas" breaks the hearts of those who have no home they value. The vision of laughing children, drowning in toys and Dickensian plenty, pains the poor children who rarely feel anything but hunger. Ask any psychiatrist who must carry his patients through the holiday season and he will tell you that for every joy of Christmas there is a fright of Christmas, too.

So Martin H. Greenberg, Charles G. Waugh, and I have gathered together in *The Twelve Frights of Christmas* a group of stories that will, in one way or other, show you the other side of the holiday season. It is not that we don't want you to have a merry Christmas and a happy New Year—we do!—but the other side exists, and nothing human should be alien to you.

# 80
# Valentine's Day

The Latin word *valere* means "to be strong," and from it we get such words as "valiant" and "valor," since one expects a strong person to be brave. We also get words such as "value" and "valid," since strength can refer not only to muscular power but also to something that finds its strength in being worth a great deal or in being true.

In naming children, we can make use of words that imply the kind of character or virtue that we hope to find or instill in him or her. This is not so true now, when we use old traditional names that had meaning in archaic times we are no longer familiar with, but we think it fitting that a cartoon hero renowned for his strength and bravery be called "Prince Valiant."

The ancient Romans, by the same reasoning, might use the name "Valens," which means "strength." By the irony of history, such a name became particularly popular in the latter days of the Empire, when Rome had grown weak. It was as though the Romans were using cheerful names to hide the dismal reality.

Thus, there was a Roman emperor named Valens, who ruled from 364 to 378, but his name turned out to be particularly inappropriate. He fought the Goths at the battle of Adrianople on August 9, 378, and his Roman legions suffered an overwhelming defeat—a defeat from which the Empire never recovered. Valens, whose poor generalship was partly responsible, was himself killed in that battle.

Most names have "diminutives," for it is almost inevitable that babies who receive a sonorous name be given a pet version of that same name, so that you have Dickie instead of Richard, Bobby instead of Robert, and so on. And sometimes the diminutive is retained into maturity. The Roman fashion of forming diminutives resulted in the name "Valentinianus," meaning "little Valens."

As a matter of fact, the Emperor Valens had an older brother who was named Valentinianus. (Yes, the older brother had the diminutive form of the name; these things don't go by logic.) Valentinianus reigned as co-Emperor and was the more capable of the two brothers. He died first, in 375. He was succeeded by his four-year-old son, who reigned as Valentinianus II till he was assassinated in 392. There was also a Valentinianus III, who reigned from 425 to 455 and was an almost complete nonentity in the days when the Western portion of the Roman Empire was tottering to its fall.

The English language tends to shorten the long Roman names, so that the various Emperors of that name are known as "Valentinian" to us. A shorter form, even to the Romans, was "Valentinus," and this is shortened in English to "Valentine."

During the days of the Roman Empire, there were periodic persecutions of the Christians, and the Christian Church celebrated the martyrs, those who were executed in the course of these persecutions. The martyrs were usually awarded sainthood, and the days on which they were executed were held sacred to their memory. In the course of time every day in the year had one or more martyrs attached to it.

The martyrs were often of Roman descent and had Roman names. At least two of them were named Valentinus (or Valentine to us), and the day held sacred to their memory is February 14, which is therefore "St. Valentine's Day."

Now it is necessary to turn to something else.

In ancient times, when infant mortality was high and life expectancy low in any case, it was important to have many children. Consequently, people who, for any reason, had few children or none considered themselves under a curse or malevolent spell and went to some lengths to ensure fertility by some religious or mystical rite.

The ancient Romans had a holy spot where (according to legend) the wolf had suckled the twin brothers, Romulus and Remus, the former of whom eventually founded Rome. The spot was called "the Lupercal," from the Latin word *lupus,* meaning "wolf."

On that spot, every February 15, there was a festival called the Lupercalia, during which animals were sacrificed. Thongs were prepared from the bloody strips of animal hide, and priests ran through the crowd striking out with those thongs. Those who were struck were considered to be cured of sterility. Naturally, those who wanted children flocked to the festival. Afterwards, I imagine, they engaged in those activities that were expected to give rise to children—striking while the iron was hot, so to speak. Consequently, the Lupercalian festivities were associated with love and sex.

In 494, Pope Gelasius I forbade this pagan festival, but that sort of thing does no good. The festival simply continues under another name. For example, the celebration of the winter solstice was forbidden, but it still continues with almost all the pagan customs of the ancient Romans—under the name of "Christmas." To the celebration of the vernal equinox was added the Christian feast of the resurrection, which became "Easter," and so on.

The Lupercalian festival of February 15 simply became St. Valentine's

Day of February 14. (The change of one day may have come about because St. Valentine was a popular saint.) Legends arose later to the effect that St. Valentine had been kindly to lovers, but that is undoubtedly just a cover for the good old fertility rites that have always been popular (and, I strongly suspect, always will be).

The day was trivialized, of course, by the greeting-card industry. Because most people are inarticulate and find it difficult to compose a letter, they gladly pay a small sum to purchase a synthetic letter. The industry encourages this, and while the first greeting card was a Valentine's Day card, manufacturers have instituted cards for everything imaginable, right down to celebrating the day on which one's second cousin first fell into a mud puddle.

Consequently, a "valentine" has come to mean such a card and, by extension, someone's sweetheart. The word is redolent of paper-lace and pastel shades and cherubs and hearts, and it is only because we are so accustomed to all that stuff that we avoid nausea.

In any case, "valentine" has become such a pleasant word that the thought of anything unpleasant taking place in connection with Valentine's Day carries a double load of shock, and it is this double load which you are invited to experience in the stories of this collection, *Fourteen Vicious Valentines.*

# 81
# Hobgoblin

I don't particularly value consistency, unless it is truly consistent.

Very often, it isn't truly consistent; it merely carries a varnish of lookalike; and to buy the nothing-content for the outside varnish is to fall prey to a foolish consistency. "A foolish consistency is the hobgoblin of little minds, adored by little statesmen and philosophers and divines," as Ralph Waldo Emerson said.

But let us not talk in abstractions and generalities; let's get down to specific examples.

I am reasonably well-known as a science-fiction writer, and in my stories, at least in the person of my characters, I whiz along through outer space, visiting the far stars of the Galaxy.

In real life, I refuse to get on airplanes.

Come the little philosophers in many shapes and forms and say to me, "Isn't it strange that in your writing you visit the outermost corners of the Universe and yet in real life you will not fly?"

It seems inconsistent, doesn't it?

Well, let's see if that's so. I compose all my fiction at my keyboard. My keyboard has never crashed; it has never been skyjacked; it has never run out of fuel; its engines have never caught fire. In what way does it compare to an airplane? Why should anything I do at the keyboard force me to take an airplane? Why must I, in reality, mimic the actions of my characters, in fancy? Where does the consistency come in?

To bind fiction to reality, my fancy to my actions, my keyboard to an airplane, is to insist on a *foolish* consistency—and thank you but I will not oblige.

If I decide to fly, I'll fly for good reasons. The fact that I write science fiction is not a good reason.

* * *

Second example, and more immediately to the point.

I'm a rationalist and I'm quite loud-mouthed about it. I stand resolutely against the faddish half-bakery of the world, and why not? One thing of which I am quite certain is that a belief has only to be silly to attract the undying devotion of those millions whose brains match that belief in quality.

Sure, I'm ready to believe that UFOs are extraterrestrial spaceships, as soon as the evidence is compelling. I'm even ready to *consider* that UFOs *might be* extraterrestrial spaceships, as soon as the evidence is no more than suggestive. In fact, I'm ready to suspect that further investigation of UFOs might be worthwhile, as soon as the evidence is at least *interesting*.

As long as all the evidence concerning UFOs is, however, dished up in dreary anecdotes that offer no handles for further checking of even the most elementary sort, I won't even take the effort to yawn at it. If such an attitude arouses the UFO devotees to fury, that makes me feel all the more comfortable, since I know that anger is the common substitute for logic among those who have no evidence for what they desperately want to believe.

The same goes for Velikovskian catastrophes, for von Dänikenish ancient astronauts, for Gellerian spoon-bending, for mental powers of all varieties, for conversations with planets, for pyramid power—as well as for the more ancient notions of ghosts, spirits, fairies, angels, demons, astrology, necromancy, witchcraft, and all the varieties of magic.

Surely, then, a rationalist like myself would *love* the hard-headed old-fashioned mystery in which all the clues are firmly on the table and in which a cold and inexorable chain of logic begins at the evidence and ends at the solution, skipping over all the red herrings and dodging round all the planted irrelevancies.

And I do, I do! I love all that logic! Isn't that consistent of me? You betcha!

And surely, then, a rationalist like myself would *hate* the fuzzy fantasies about ghosts and spirits and curses and voodoo and telepathy and ominous dreams and "second sight" and all the rest of that silly nonsense that has plagued the silly and nonsensical all through human history.

But I don't, I don't. I love all that fuzzy fantasy, too! I love to feel the cold chills and to shiver at the hint of things beyond and all that scary stuff.

Inconsistent? Me? Hell, no. If I rejected fantasy in fiction because I reject it in real life, I would be confusing fiction and reality and would be indulging in a foolish consistency, which is not my thing at all.

Consider! Why do I reject fantasy in real life? Because I accept the two basic assumptions that underlie the scientific view of the Universe: (1) that the Universe runs in accordance with a few very general and very powerful basic rules, called the "laws of nature," which do not change and cannot be subverted, and (2) that it is possible for the human mind slowly to work out those rules and interpret reality in their light.

Those are only assumptions and therefore can't be proved. I accept them on faith, and in that sense the scientific view of the Universe is my religion.

However, I accept them for a reason. A Universe that is under the rule of Law and that is comprehensible is a likable Universe, one that I find warm and comfortable to live in.

To be sure, the nature of the Law is as yet imperfectly understood, and little by little it is being more deeply and broadly understood. Every inch of comprehensibility must be fought for with every atom of our mind—but that makes it a fun Universe and one that I find exciting to live in.

Accept fantasy, though, and you have a Universe run by the whim of incomprehensible gods or demons. The Universe becomes a terrifying place, in which ignorance is raised to the supreme virtue and blind obedience to the supreme act. Nothing is left for human beings to do but to fawn and beg. Not for me, thank you.

But what has all that to do with fiction? Must one refuse to read accounts of the climbing of Mount Everest because one would refuse on any account to make the attempt one's self?

What pleasure it is to enter into the terrifying world of fantasy for a few moments and to live, vicariously, the kind of life in which the rules can be broken, where logic might be no sword and rationality no shield?—Just so long as it's not the real world, and you *know* it's not the real world.

There is no inconsistency in enjoying a temporary entrance into a world that you know doesn't exist—but fighting to the death any claim that it does exist.

And to my own taste, the most piquant flavor arrives when you mingle the opposites. If you can take the classic mystery, with its super-rationalisms, its creation of a small sub-Universe in which all the evidence is there and the logic is flawless, its evocation of a world in which there isn't even the real-life fuzziness of the incomplete—and then add to it not merely the incompleteness of the real, but the lawlessness of the fantasy, you've *got* to enjoy it.

Well, *I've* got to enjoy it, anyway, so I love this collection of piquancy and spice: *Mysterious Visions.*

# 82

# All the Ways Things Can't Happen

You don't have to be very old before you get it quite through your head that things can only happen in certain ways. If you throw a ball in the air, it's going to come down again, and once you eat a piece of cake, you haven't got it any more.

People even make up sayings about it: Everything that goes up must come down. You can't have your cake and eat it, too.

That's all very well. It's convenient to know what the world has in store for you, so that you don't get too many nasty surprises. You wouldn't want to be walking down the street, minding your own business, and suddenly find yourself falling upward into the sky with nothing to stop your fall.

Just the same, provided you're really quite safe, it's fun to think about things that can't happen. Or, if your own imagination runs out, it's nice to read stories about it. What if anything you wished for came true? What if your dog could talk? What if you had a tame dragon and he was willing to take you for a ride?

That's what fantasy deals with. Fantasy is what happens when the "laws of nature" no longer rule the world. Fantasy pretends that things happen in all the ways things *can't* happen.

One of the advantages of fantasy is that there is such variety to it. That's because there are so many more wrong answers than there are right answers. Thus, if you're considering something like 8 × 3 + 4, there's only one right answer, 28. There are, however, an infinite number of wrong answers. For instance, 29 is a wrong answer; so is 2,000,000,000,000; so is 1½; and so is 15.179.

In the same way, laws of nature allow things to happen in only one way. You throw a ball, up and it comes down. That's it.

In fantasy, a ball, once tossed into the air, might keep on going up forever; or it might come to a halt two feet over your head and move along with you wherever you go; or it might turn into a bird; or it might simply remain motionless until you shouted "home!" and then it would lead you straight home even when you were lost and didn't know where you were. And so on.

Myths and fairy tales deal with all kinds of odd creatures: centaurs, sphinxes, basilisks, demons, and so on. The ancients must have thought such things existed, but we know they don't.

Oddly enough, knowing these imaginary creatures don't exist makes it all the more interesting to read about them. They make the world stranger and more full of wonders. In this book, *Isaac Asimov's Adventures in Fantasy,* you can read a story about a centaur and two about evil spirits.

Even when some creatures *do* exist, they sometimes only exist in far-off places. If you live in Kansas, you've probably never seen an okapi or an echidna. If you live in Australia you might never have seen a moose or an anaconda. Such animals are not fantasies, however. If you go to the right place, you will see them.

But what if you saw a whale, a full-size huge whale, in a swimming pool in Arizona. *That* would be fantasy. In this book there's a story about such a whale, and about something along with it that is even more unusual.

Or what if you could turn yourself into an animal if you liked? What animal would you choose? For myself I would like to be an Andean condor for just an hour, so that I could know what it was like to soar in the air a mile high and feel the wind in my feathers and see the world spread out like a map beneath me, and then dive and climb again.

# 83

# Is Fantasy Forever?

In some ways, all fiction writing is fantasy. If a tale is truly fiction, it never happened; and if it never happened, it is fantasy; it is a creation of the mind, the imagination. For that matter, if we want to be very strict about it, much supposed "nonfiction" is fantasy, too.

The fact is, though, we *don't* want to be very strict about it. If we define fantasy in such a way as to include almost everything, then the word loses its force and it comes to mean no more than "writing."

Let us look for a different definition. Fantasy should mean not only something that is not so (and therefore exists only as an idea), but something that *could not possibly be so* and therefore can exist *in no other way* than as an idea.

Thus, Charles Dickens's *Nicholas Nickleby* is not a fantasy. Although its characters never existed and its events never took place, those characters and events could have existed without upsetting the accepted order of the Universe.

On the other hand, Dickens's "A Christmas Carol" is clearly a fantasy, for it deals with ghosts and with abstractions, like "Christmas Past," that have been made concrete. The accepted order of the Universe does not include ghosts and concretized abstractions.

In fact, we can be stricter still and insist that fantasy deal not only wtih matters that we conceive as not capable of existence in our Universe, but which we insist are incapable of existence even in a Universe modified by reasonable scientific advance. If reasonable scientific advance *could* make them possible, then we would have science fiction. (To be sure, an ingenious person can manipulate the possibilities of scientific advance in such a way that what we would casually think of as fantasy can be made into a kind of science fiction. Usually, however, the manipulation is not bothered with, so that fantasy and science fiction remain distinct.)

And now that we have an idea as to what we mean by fantasy as a restricted branch of literature, we have a right to ask how old it is. It might seem a fair guess that fantasy is forever, that it is as old as language, as old as the human imagination.

It would seem that over the Stone Age campfires our uncivilized ancestors froze each other's blood with tales of monsters, and ghosts, and demons of all sorts.

We'll never know that for sure, of course. So, if we prefer to cling to greater certainties, we have to turn to the oldest surviving scraps of literature, and these, we find, are quite likely to deal with fantasy.

The *Epic of Gilgamesh,* written by nameless Sumerians about 2700 B.C. is, I believe, the oldest surviving work of fiction, and it contains elements of fantasy—gods, monsters, plants that confer immortality, and so on. The *Iliad* and *Odyssey* are to some extent fantasies, especially the latter. The tales of Polyphemus the Cyclops and of Circe the Witch remain, to this day, among the most popular fantasies in existence.

Folktales are almost invariably fantasies; the stories in *The Arabian Nights* are fantasies, for instance, as are *Snow White* and *Cinderella.* Every age has its fantasies, and even the twentieth century has developed some that rival those of the past in skill and popularity. Consider *Mary Poppins, The Hobbit,* and *Watership Down.*

And yet—when is a fantasy not a fantasy?

The answer, surely, is this: When its events are not accepted as running contrary to the accepted order of the Universe. Even more so, when its events, however fantastic they may seem, are accepted as literal truth.

Thus, the Bible is filled with wonder tales: the speaking serpent in the Garden of Eden, the speaking ass that Balaam bestrode, the parting of the Red Sea, the deeds of Elijah and Elisha, the activities of Jesus as a healer. If these were encountered by some well-educated Chinese person who had never heard of the Bible before, he or she would have no hesitation in labeling the book a fantasy collection. Naturally, pious Jews and Christians would reject such a view with horror and would consider it blasphemous.

In the same way, unsophisticated people of the past who believed in the Olympian gods and goddesses, and who had no doubt that strange monsters existed in the misty regions beyond the small patch of ground they knew well, would accept Homer's tales as accurate history in all its details.

And in later times, those who believed in ghosts, or afreets, or ghouls, or fairies, or elves, would accept tales involving them as at least true in concept, if not necessarily in detail, and they would not be thought of as fantasies at all.

How far into the present does this notion of "fantasies that are not fantasies" extend? Obviously, right into the present and, probably, into the future as far as the mind can see. Every religion seems like a fantasy to outsiders, but as holy truth to those of the faith. There are always people who are unsophisticated, because of youth or lack of modern secular and scientific education, who believe in Santa Claus, in zombies and voodoo, in the Tooth Fairy and the Easter Bunny, and so on.

There are even adults who, to all appearances, are intelligent, educated, and sophisticated, who are nevertheless believers in astrology, spiritualism, creation-science, or other irrationalities, which seem like nonsensical fantasy to those of us who are untainted by such things.

In that case, when, if ever, did we start thinking of fantasy as *fantasy?*

No doubt there were always some skeptics, some people we would view today as hard-headed realists, in even the most superstitious and faith-ridden times. These people scorned anything not based on observational and rational evidence, and they were firm in the belief that what most people accepted without question was, in actuality, mere fantasy.

This is, however, not enough. The occasional skeptic can barely make a mark on society. Did there come a time, however, when such rationalism became an accepted part of a secular society and when people in reasonably large numbers were educated into the belief that the Universe could be understood only by reason, so that anything beyond that was fantasy? Such a state of affairs began to arrive in the Western world after the end of the period of the religious wars with the coming of the Age of Reason. The latter half of the seventeenth century, the time of the Royal Society and of Isaac Newton, marks the dividing line.

Even then, however, rationality was confined to a rather thin layer of the educated. It was not till the nineteenth century that, in the Western world, there gradually rose the notion of mass-education under the control of a secular state. For the first time, there were large regions in which high percentages of the population were educated in school systems that were not run by some religious group or other. And then, for the first time, there rose large numbers of individuals who could tell what fantasy was, and who enjoyed it all the more because they recognized it as pure exercises of untrammeled imagination.

I have said elsewhere that the nineteenth century was the first century of science fiction. I believe, for the reasons I have presented, that it was the first century of true fantasy as well, and the stories represent in *Isaac Asimov Presents the Best Fantasy of the Nineteenth Century* are the prize examples of that first century.

# 84

# Wishing Will Make It So

When I was much younger than I am now, I heard the philosophical comment: "It takes a million dollars to make a millionaire, but a pauper can be poor without a cent."

At another time, I listened to Sid Caesar playing the role of a Teutonic mountaineer. Carl Reiner said to him, "Tell me, Professor, how long does it take a person to negotiate the distance between the top and bottom of a mountain?"

Said Sid, "Two minutes."

Carl said, with considerable astonishment, "It takes only two minutes to climb a mountain?"

To which Sid said, with disgust, "Not climb. To negotiate the distance from the top down to the bottom—two minutes. *Climbing* is a different thing altogether."

I've thought about such things, and it became clear to me that both the examples I have given are representative of a general state of affairs that can best be expressed as follows: "Lousy things are no trouble."

For instance, it's no trouble to go hungry. You don't need money, and you don't have to make an effort. You just sit there. Getting yourself a square meal can be very troublesome, however.

Again, suppose that someone brings you all the food you can eat. In that case, it's getting fat that requires no effort (if you don't count the tiny effort it takes to lift the food to your mouth, chew, and swallow). To avoid fat, however, means eating less than you probably want to, and engaging in vigorous exercise besides.

This is not something that has escaped the notice of humanity. I'm absolutely certain that even the meanest intelligence has noticed how readily one can be poor, hungry, thirsty, cold in the winter, hot in the summer, while finding oneself with nothing to wear, nothing to read, and nothing pleasant to do.

Not only does one have to take trouble and make an effort in order to avoid all these lousy things for which there are no charge, but there is no limit on the quantity of trouble and effort you may have to make. Most people can work hard all their lives, and stint no effort doing so, and yet find themselves far short of the millionaire mark when they're through.

You may want to marry a rich man's gorgeous daughter (or, if you are a woman, his handsome son), and for that purpose you may bring into play every bit of charm you have—and get nowhere. This can start you brooding over the fact that you can probably, without any effort at all, succeed in marrying any number of very poor, very ugly women (or men).

Well, then, what are you going to do? You crave pleasant things, which take more of an effort than you can possibly pump up in a lifetime of pumping, and you want to avoid unpleasant things that are being forced upon you against your will and that then stick to you despite your shouts of dismay.

It is easy to decide that there is something wrong with this. In a properly run Universe, surely you deserve to get something simply because you want it. Even though this doesn't *seem* to happen, there must surely be some trick to bring it about. Perhaps there is some formula or spell that will give you anything you want; you need only wish for it. Or else, perhaps there is some supernatual being willing to gratify you under certain conditions. Perhaps there is some wishing object that already exists, manufactured who knows how, that you need only find in order to gratify your every wish.

Folklore of every kind includes tales of magic wishes, and the most successful of all such stories is to be found in *The Thousand and One Nights* (more commonly known as *The Arabian Nights*). What child isn't fascinated by the tale of "Aladdin and his Lamp" and doesn't fantasize having such a lamp? I experienced both the fascination and the fantasy in copious quantites when I was young.

(Incidentally, we moderns still believe in the power of wishing. We call it "praying," of course, and, all too frequently, praying is simply a way of substituting God for the Slave of the Lamp and making him run our errands for us.)

Of course, some such tales caution against overweening greed. Midas, having wished that everything he touched turn to gold, found he had gone too far, leaving himself no way of eating or drinking, so that he had to beg to get the wish canceled.

In other stories the wishes are limited in number, most often to three, and then, invariably, there is a problem in deciding what the wishes ought to be. Almost as invariably, the choices prove unfortunate.

This instinctive suspicion that the notion that "wishing will make it so" is nonsense was given its final support by the laws of thermodynamics. The first law says that the amount of energy in any system is limited and the second says (in scientific terms) exactly what I said earlier—that lousy things are no trouble, but that to accomplish anything desirable takes an effort. What's more, the laws of thermodynamics hold for everything in the Universe, including Slaves of the Lamp.

And yet— And yet—

Even if we are grown-up, hard-headed, scientific, and have put childish things behind us, there is still this hankering. Even though we know that wishing will not make it so, we can't help but *wish* that wishing will make it so.

Here, then, are seventeen stories in which wishes, in one way or another, are involved. And just to make sure that you will be hooked by them, the first story, "The Monkey's Paw," is, to my way of thinking, the best such story ever written, and the grisliest. How I envy you, if you're now reading it for the first time.

So suspend your disbelief for a while and enjoy *Magic Wishes.*

# 85

# Wizards

There is nothing really mysterious about the word *wizard.* The first syllable, *wiz,* is used in common slang these days for anyone or anything that is uncommonly smart or impressive, and it sounds very much like *wise.* In fact, it *is* a form of "wise," and a "wizard" is simply a "wise man."

Of course there's the suffix *-ard,* which, along with its variant *-art,* is usually used to indicate an excessive amount of something. A "coward" is one who is too easily cowed; a "braggart" is one who brags entirely too much; a "drunkard" is one who is too frequently drunk.

And a "wizard," then. Presumably, a wizard is one who is too wise for anyone's comfort.

How is that possible? In general, we tend to honor wisdom, to deify it almost. How can one be too wise?

It depends on the kind of Universe we live in. To almost all people in every generation—even our own—the Universe is a frightening and dangerous place. It is operated on an entirely whimsical, capricious, and even clearly malevolent basis, and we are the helpless prey of forces enormously greater than we can handle.

How else can we explain the storms that strike without warning, the droughts, the sudden onset of disease and plagues, the mischances of every kind?

Surely, the Universe must be under the control of beings who are as irrational, as erratic, and as irascible as human beings are at their worst; who are incredibly powerful and yet incredibly childish as well; who, even if basically well-disposed, are apt to explode into uncontrollable anger at some small offense or meaningless slight.

Even if we picture the Universe as under the control of an all-good, as well as all-powerful, being, that being is apt to lose his or her temper and then—watch out. Or, if the being is so good that his or her goodness is never for an instant in question, one can only suppose the additional existence of competing forces of evil, which the all-good being is forced to allow to exist (or chooses, for some inscrutable reason of his or her own, to allow to exist).

However we may slice it, the Universe seems to be a horrible madhouse. Yet might it not be possible to behave in such a way as to keep all these supernatural beings in good temper? You might kill animals and burn them so that a delicious smoke rises to the sky where these beings live and feed them into good-natured satiation. Or you might sing endless songs of praise to these beings, flattering them into benevolence. Or you might find magic chants that either lull them to kindness or bind them into impotence.

Everything, however, must be done *just so.* The words, the gestures, the exact order of events, the whole ritual, must be correct, or the result will surely be worse than if you had done nothing at all.

But how do you discover what the ritual must be? Clearly, the only sources are the supernatural beings themselves. If some human being learns the secrets, he or she can control the Universe by flattering, bamboozling, or overpowering the supernatural beings.

Does any human being actually learn the secret? Well, you and I, being very clever people who live in the twentieth century and who have had an excellent scientific education, know that they don't, that there are no secrets of this sort, that there are, indeed, no supernatural beings of this sort, no demons, afreets, jinns, nymphs, and satyrs; but that's just you and I. To others not so well placed in space and time, and not so learned and sophisticated as we are, it is enough if someone *says* he or she has the secrets. If they are clever enough and daring enough to make those who watch and listen think they are indeed controlling the Universe, they will believe them. (Why not? Think how many millions fall for all the hoary old tricks and flim-flams form astrology to spoon-bending, plus everything in between.)

There are various names one can give the people who know the secrets whereby the Universe might be controlled, but one of them is "wizard."

People might feel grateful to the controllers of fate, for surely it is to them that one must turn to make sure that the rains come and the infections don't; they are the saviors, the answerers of questions, the bringers of good fortune, the helpers in times of disappointment and sickness.

Think of Merlin, the archetypal wizard of legend and perhaps the most popular of all. Who has a bad word for him?

Then why that "-ard" ending?

Is not a wizard just as capable of irascibility and loss of temper as any supernatural being? Might not a wizard have his or her feelings hurt? Might he or she not hunger for more power? In short, are not wizards just as dangerous as the beings they control?

Of course.

Wizardry is a double-edged sword, then, and if we deal with stories of wizards, which edge are we likely to harp on? Remember that catastrophe is more dramatic than peace, danger more dramatic than calmness, and—yes—evil is more dramatic than good. Writers, being human, and wanting to go where the readers are, are apt, therefore, to stress the evils and dangers of wizardry.

What you have in the book *Wizards,* then, is a group of stories that are full of drama, danger, and heart-stopping action. That's the best kind of stories to have, as long as you're sitting comfortably in your favorite armchair, or cuddled cozily in bed—so best wishes to you, and good reading.

# 86

# Witches

In Anglo-Saxon times, a magician was called a *wicca.* At least, a male magician was. A female magician was a *wicce,* the feminine form of the word. In modern English, word became "witch." You might think that we would use the modern English system for deriving a female from a male noun, and speak of a male magician as a "witch" and a female magician as a "witchess." After all, we talk of "enchanters" and "enchantresses," of "sorcerers" and "sorceresses."

But it doesn't work for "witch." That term is always used for a female magician. We simply don't think of males as "witches."

On the other hand, a "wizard" is always a male magician, and no one uses the term "wizardess."

For that reason, in modern English we think of "wizard" and "witch" as the male and female version, respectively, of the same word. After all, we don't have to have every female word formed by adding the "-ess" suffix. We may have "prince/princess" and "count/countess," but we don't have "king/kingess." Instead, we have "king/queen" and, for that matter, "gentleman/lady," "boy/girl," "bull/cow." Why not, then, "wizard/witch"?

You can use each word in the original meaning of magician or in a modern diluted meaning. A wizard can be merely some man who is extraordinarily clever or good at his task. A witch can be merely some woman who is so roguishly lovely that men cannot resist her. Both are so good at what they do that they seem to use magic.

Yet there is an important difference, too, an asymmetry. If we consider the words in their original meaning—of someone who makes use of magic to accomplish some aim—that aim might be either a good or an evil one, in intention and in fact. In the case of a wizard, it is liable to be good (to our casual thinking) as often as bad. The greatest wizard in literature is the one in the Arthurian legends—Merlin. And he is certainly considered to be on the side of good. The most popular wizard in modern children's literature is the Wizard of Oz, and he is also on the side of good.

Witches, too, might be either good or evil, and in the Oz books Glinda is a good witch. However, that is quite exceptional. We almost always think of witches as evil. In the movie *The Wizard of Oz* we tend to forget the pale, washed-out Billie Burke as the wholly improbable good witch, and we remember only the magificent performance of Margaret Hamilton as the Wicked Witch of the West. In literature generally, the most memorable witches are the three weird sisters in *Macbeth,* and they, too, are clearly evil phenomena.

Why, then, are witches so much more likely to be viewed as evil than wizards are? There, the fault might lie with the Bible. Witches, wizards, and all varieties of enchanters and magicians were thought to use the powers of demons and devils, and therefore to be working in opposition to God, so that they were considered to be evil virtually as a matter of course. Their tampering with the supernatural was held to be a capital offense in the Bible. The Hebrew word for such tamperers was translated into the English of the King James Bible as *witch,* so that we have the verse: "Thou shalt not suffer a witch to live" (Exod. 22:18). A word equivalent to "witch" was used in the translation of that verse into other European languages.

As a result, during the first century or so after the Protestant Reformation, when Europe was torn by religious wars, and everyone was tense over religious differences, a virtual witch-hunting mania swept over Europe, and thousands of innocent people were tortured and killed. Although some men suffered, the vast majority were women.

Even though educated people no longer believe in witches (in the sense of thinking that there are people who can truly manipulate supernatural entities), the aura of evil remains in connection with witches rather than with wizards.

A woman may be a witch at any age. After all, what's to stop a young woman from dealing with demons? (Thus, in this anthology of stories about witches, there are young as well as old witches.)

Nevertheless, when we think of witches, we generally think of old women. What's more, in the sad days when witches were being hunted down by the hundreds, old women suffered in numbers far out of proportion to their occurrence in the general population.

The witches in *Macbeth* are always pictured as old women; the traditional witches with which we are presented in Halloween illustrations are always old women. Why old?

Here's what I think:

Up to about a century and a half ago, the average life expectancy was only thirty-five at the most, and old people made up a considerably smaller percentage of the population than they do now. What's more, aside from the causes of death that dealt with either sex indiscriminantly, women frequently died of childbirth, too. For that reason, there were fewer old women than old men.

The face changes with age. For one thing, it becomes wrinkled and withered. For another, teeth disappear. In the days when teeth weren't cleaned and were certainly never cared for properly, it was probably a rare human being who had any teeth left after the age of forty.

In the case of old men, the wrinkled face and the toothless gums were hidden by the beard that almost all men wore. Old women, however, had no beards, and the few who existed must have looked very odd, and even frightening, with their faces so different from those of younger people. And since difference is often equated with evil, old women were easy to think of as evil.

Look at the old witches of Halloween. Invariably, they have pointed noses and chins that approach each other. That is virtually the trademark of the witch. However, when there are no teeth in the mouth, the nose and chin naturally approach each other more closely than when teeth serve as a barrier. The Halloween witch is merely a toothless old woman.

I wonder if modern dentistry might not have done more to wipe out fear of witches than any amount of education. That and the fact that with the general extension of the lifespan, old people—and old women, in particular—have grown too common to scare anyone.

But in stories, they still exist, and they can still frighten. So welcome to *Witches.*

# 87

# Curses!

If you have an enemy you are desperately anxious to get rid of, the most straightforward thing to do is to get a big club and bash him or her over the head. That's it!

But what if he or she is bigger and stronger than you are? Or what if he or she has a big family that will get after you once you've done your job? Or what if you live in a society that has a large and interfering police department that will then pursue and arrest you?

You're stuck with the next best thing. You must confine yourself to *wishing* your enemy were dead. Naturally, if you believe in the existence of gods or demons, and if one of them happens to be a friend of yours, perhaps you can persuade said god or demon to do the job for you. You would use some magic chant, or perform some magic ritual, or both, in order to cajole or force the god or demon to oblige. You must, in other words, cast a "magic spell"—where "spell" is an old Teutonic word for "chant" or "tale." (Thus, "gospel" is actually "God's spell," "the story of God.")

Of course, the magic spell doesn't have to bring *death.* Death is, after all, not much of a punishment. A moment of dizziness, a pang or two, the coming of blackness and then—nothingness. What kind of punishment is that? Most people find that rather dull and unimaginative, and definitely unsatisfying. That is why a lot of thought has gone into inventing and describing Hell (or its equivalent) as a place where people you don't like can be consigned after death, so they can be tortured in horrible ways not just for a rotten trillion years but *forever!*

Or else, the victim of a magic spell can be left alive in order to suffer something that will make them feel a lot worse than just dying will. For instance, suppose you order a man and a woman not to eat a certain kind of fruit, because if they do you will kill them. "In the day thou eatest thereof thou shalt surely die," you say, solemnly.

And then the man and woman go ahead and eat the fruit, and you decide that for a serious infraction of the rules like that, death isn't enough. So, since you're omnipotent, you break your solemn word and say to the man, "In the sweat of thy face shalt thou eat bread," and to the woman you say, "In sorrow thou shalt bring forth children." In other words, the man is condemned to lifelong hard labor and the woman to the repeated pangs of childbirth. What's more, all the descendants of these two are condemned to the same punishment!

Isn't a magic spell like that much better and more satisfying than just bashing people over the head?

It's no surprise at all, then, that throughout history magic spells have delighted humanity. It got so that nobody could die of disease, or have an accident, without everyone else in the community suspecting that a magic spell was the case. After all, what else could possibly bring about such things? And as though that weren't fun in itself, there would then be the additional pleasure of finding someone you don't like—an ugly, old woman who's no good to anyone is the best bet—accusing her of casting the spell, and torturing her to death.

Those were the good old days.

In theory, to be sure, magic spells don't have to bring misfortune. They don't have to be "curses," in other words. They might be used as "blessings" to ensure good fortune, or to neutralize misfortune. Such things are indeed sometimes described, but very little interest is aroused in them.

I suppose the psychology of that is not very difficult to understand. If things are going well with you, there's no need to feel that magic is involved. After all, things *ought* to be going well with you. You're a wonderful person, as you yourself will testify (and who knows you better than you do?), and wonderful things should naturally happen to you. And as for wishing good fortune on someone else and blessing them, what the devil have *they* done to deserve it?

Thus, in the Gilbert and Sullivan opera *The Sorcerer,* John Wellington Wells (who is the sorcerer of the title) is describing his stock in trade: "We have some very superior Blessings, too," he says, "but they're very little asked for. We've only sold one since Christmas. . . . But our sale of penny Curses, especially on Saturday nights, is tremendous. We can't turn 'em out fast enough."

There you are—would William Schwenk Gilbert lie?

So popular were magic spells, in fact, that even though they were seen everywhere in real life, that wasn't enough. They penetrated literature as well. Think of all the folktales populated by wicked witches, and offended fairies, and haughty enchanters, and devious sorcerers, all of them with an inexhaustible fund of curses to hurl at those unfortunates who have committed such terrible crimes as forgetting to invite them to a christening.

Of course, in our present disgustingly secularized times, when humanists and such-like scum abound, and when belief in gods and demons and magic spells (whether blessings or curses) are frowned upon by vicious modernists, there is a sad falling off in the good old customs. Witches are hardly ever burned any more (though there's a move on among the pious, I understand, to burn liberal Democrats who refuse to vote right).

Fortunately, however, magic spells and the hearty curses of yore live on in our modern fantasies, and just to bring you all a breath of fresh air, we are supplying you with a dozen or so stories of this kind for your pleasure and delectation.

What's more, the fact that Martin H. Greenberg, Charles G. Waugh, and I have edited *Magic Spells* gives us certain powers. It is only fair to warn the critics, therefore, that if any of them are so short-sighted as to say bad things about this book, their word-processors will go down at the first stroke of a key. And if they use a typewriter, a toad will leap out of their machine each time they complete a sentence.

(At this point I wanted to describe the blessings that would be showered upon them if they gave us a good review, but Martin and Charles insist that critics don't understand kindness and can only be dealt with severely.)

# 88

# The Forces of Evil

Halloween is the time we confront the forces of evil, when devils, imps, witches, and goblins are on the prowl, and that seems odd if you stop to think about it.

November 1 is All Saints Day. Each saint has a day of his or her own—the day of his or her martyrdom, or of some other salient event in his or her life—but on All Saints Day *all* the saints are celebrated. An older term is All Hallows Day, since both "Hallow" and "Saint" are derived from the notion of "holiness," of "sanctification," of "devotion to the service of God"; the former word came from the Teutonic, the latter from Latin.

Among the ancient peoples (and among Jews even today), the day was supposed to start at sunset. Thus, by ancient standards, Christmas would start at sunset on December 24. That is why we make a fuss about "Christmas Eve." It is not just the evening before Christmas; it was originally the first part of Christmas itself. The same is true for "New Year's Eve."

Similarly, All Hallows Day should start, by the old tradition, at sunset on October 31. That evening would be All Hallows Eve, or All Hallows Even if we use the older name. All Hallows Even is easily abbreviated to "Hallow Even" and, further, to "Halloween."

But then, how is a day that is devoted to all the saints, to all that is holy, come to be celebrated as a day when all the forces of evil are broad? In fact, where do the forces of evil come from?

The ancients always recognized that both good and evil were abroad in the world, but the mythological treatment of this varied from culture to culture. The Greeks, for instance, tended to think that the gods were basically good, but that they could be angered; and, when angered, they could subject humanity to evil. Thus, Apollo, the most attractive of the Greek gods, was not only the source of youth, male beauty, sunlight, poetry, and medicine—but also of disease and plagues. When he was angry, the twanging of his bow shot down men, women, and children by the thousands.

In the Norse myths, on the other hand, there was a clearer separation of good and evil. The gods (Odin, Thor, and the rest), who were basically good, faced the eternal enmity of the evil giants, and between them was an endless war, which was presumably mirrored in the presence of both good and evil in the world. Even within the ranks of the gods there was evil, for Loki, the cleverest of the gods, was more spiteful and malicious than any giant.

In this respect, however, it was the Persian mythology that was most influential in the development of Halloween. Surprised? Well—around 580 B.C. Zarathustra (Zoroaster) systematized the Persian dualistic view of the Universe. There was a principle of good, Ahura-Mazda (or Ormuzd), and a principle of evil, Ahriman, which were viewed as virtually independent of each other and very nearly equal. The creation, development, and history of the world were all incidents in the unending celestial warfare between these two principles, each leading a separate army of innumerable spirits. So even was the contest that it was necessary for human beings to choose sides, for even their puny power might sway the victory to one side or the other. (Naturally, it was to be hoped that as many human beings as possible would choose the good.)

The Jews were part of the Persian Empire for two centuries, from 538 B.C. to 330 B.C., and in that time certain Persian notions penetrated their thinking.

Before the Persians took over, the Jews had thought of God as all in all, as the author of evil as well as good. Thus in 2 Samuel 24:1 the Bible says, "The anger of the Lord was kindled against Israel, and he moved David against them to say, Go, number Israel and Judah." David's census was apparently a sin (the Bible doesn't say why), and, as a result, God afflicted Israel with a severe plague. God was both bringing about and punishing the sin.

Once the Persians came in, however, the Jews, while retaining the thought that God was supreme and could not be permanently defeated or thwarted, created a formidable adversary who could cause him at least temporary trouble. This Judaistic Ahriman was named "Satan" (a Hebrew word meaning "adversary").

In the First Book of Chronicles (21:1), a retelling of Israel's history by people writing during or after the Persian domination, the incident of the census is told as follows: "And Satan stood up against Israel, and provoked David to number Israel."

It is now Satan's fault, not God's.

As in the case of the Persian mythology, God and Satan were each at the head of an army of innumerable spirits, and by Roman times the popular religion was full of such spirits. The New Testament has a number of tales of evil spirits being cast out of human beings.

During and after Roman times, Christianity, which began as a Jewish sect, and which borrowed Jewish thinking on the subject of God, Satan, and their respective armies of spirits, slowly converted first the Roman Empire and then the rest of Europe. In the process, they encountered pagan gods. The Christians could see these only as evil spirits who, led by Satan, were masquerading as gods and misleading the people who followed them.

In Celtic Europe, notably in the British Isles, November 1 was considered the beginning of the year. The harvest was safely in and the winter's food supply was assured, so the people could relax and enjoy themselves. They thanked their gods, who, it was assumed, swarmed over the world with equal happiness.

To the Christians the Celts were invoking evil spirits, but it was difficult to argue against such a joyous holiday. The Christians therefore took it over and called it "All Saints Day"—all the holy saints and not all the evil spirits.

However, some of the people did not forget the joyous rites of the old Celtic rites, and these have survived after fifteen hundred years or so in the tiny ritual of trick-or-treat and in the surviving stories of witches and goblins (the Christian view of the Celtic deity).

Nor is October 31 the only Halloween. May 1 was also a joyous day in pagan Europe. The fields were verdant, the weather was warm, and the Sun was bright. The long winter was forgotten. With joy, people celebrated the time with fun and games (sometimes sexual), and we still have dim memories of it in the form of the Maypole and the "Queen of the May."

Again the Christians saw this as a worship of evil demons. May 1 is St. Walpurgis' (or Walburga's) Day, she being a saint, and the evening of April 30 is St. Walpurgis' Eve or, in German, "Walpurgisnacht." The German version of Halloween occurs then, with devils and witches meeting on the mountain peak of the Brocken (something that is one of the highlights in Walt Disney's *Fantasia*).

Halloween inspires these kinds of stories: stories in which the atmosphere of Halloween heightens the natural suspense already present; fantasy stories that are rooted in the witches, goblins, and devils that are inseparable from the celebration; and horror stories that take advantage of the effluvium of evil that clings to this day.

In *The Thirteen Horrors of Halloween,* which I helped edit, you will find examples of all three.

# 89
# Monsters

The word "monster" comes from the Latin *monere,* meaning "to warn." The ancients thought that any unusual happening was a warning from the gods. This was especially true if the unusual happening involved a life-form—let us say, if a two-headed calf were born.

As a result, a monster came to mean any abnormal or misshapen creature. Since human beings, viewing events, had every right to expect that such a "warning" would foretell disaster, the word "monster" was further narrowed down to any living thing that was abnormal, misshapen—and dangerous. Since large animals are bound to be more dangerous than small ones, a monster was, more often than not, expected to be very large.

It is not surprising, then, that the adjective "monstrous" came to refer to something that was enormously evil, or enormously large, or both—as, for instance, America's monstrous national debt.

When one wants to make up a good story of heroism in the face of danger, what is better than to have a reasonably normal man fight against a monster? The easiest way to construct a monster is to think of something that is dangerous in itself and then imagine it to be much larger than normal.

Hercules' first labor, for example, was that of slaying the Nemean lion, which was pictured as a monstrously large one. Simple size can be dramatic enough. The most successful monster in modern visual fiction is King Kong, who is merely a gorilla of enormous size. The sci-fi abominations of the 1950s abounded in giant spiders, giant crabs, giant dinosaurs, even giant women.

It is, however, a poor imagination that can only work up enlarged dimensions. Fantasies therefore arose in which monsters were enhanced by adding to their ugliness, their malignancy, or their powers. It was not enough to have mere giant men; one had to have ogres, who were cannibalistic, or cyclopes, who were one-eyed ogres, and so on.

A giant crocodile wasn't enough. Add wings and it becomes a dragon. Or if you want more than a venomous snake, have it breathe fire; if you have it kill not by a bite, or even by a breath, but by a mere glance, then you have a basilisk or cockatrice.

The octopus can become Medusa with snakes for hair or Scylla with six dog-headed necks, or a Hydra that can grow two heads for every one cut off. Or you can combine creatures. A man's torso on a horse's body is a centaur; a woman's torso on a lion's body is a sphinx; an eagle's body on a lion's torso is a gryphon.

Such things come under the heading of "fantasy," however. No one with even a small degree of sophistication will think that a Medusa or a Cyclops or a sphinx literally exists. But can we imagine monsters that *don't* offend our modern sense of what is reasonable and possible? If we can, we will end up not with fantasy-monsters but with science-fiction monsters.

After all, some large and spectacular organisms were discovered by Europeans only comparatively recently. For instance, no European ever saw a living gorilla until the 1850s. The initial descriptions of its size and ferocity were greatly exaggerated, and those exaggerations still live in popular thought—and lent considerable verisimilitude and plausibility to that exciting fantasy of *King Kong,* which I mentioned earlier.

Well then, might not there be other large primates *outside* of Africa? A "yeti" (or "abominable snowman") in the Himalayas, a "sasquatch" (or "bigfoot") in the northwestern United States? The chances of this are virtually zero, but perhaps that's what people like me would have said about gorillas in 1845. So a story about the abominable snowman, as a primate, perhaps even a hominid, would be considered science fiction rather than fantasy.

Then again, there is the monstrous sea serpent. Actually, there is a kind of sea serpent that exists. Discovered in the nineteenth century, it is the giant squid, which lives at moderate depths in the ocean and has tentacles that stretch out to give it the length of a whale. Perhaps there are other large creatures in the ocean depths that we still know nothing about. It was only in the 1930s that we first discovered the coelacanth, which we were sure had become extinct when the dinosaurs trod the Earth.

We might therefore consider large sea creatures to be science-fiction monsters—even a plesiosaur in Loch Ness, for which I consider the chances to be about zero. (For that matter, some even talk about the possiblity of a brontosaurus in the African rain forest—again about zero, in my opinion.)

It might even be possible to imagine a monstrous form of life starting from nonlife here on Earth and creating a horror where none existed before. We have a story of that sort in this book.

Unquestionably, though, the best monsters that can come under the heading of science fiction are creatures from other planets.

Up into the 1950s, it seemed possible that we might find extraterrestrial monsters within our own planetary system, and we have stories of *that* sort in the book, too. In the last third of a century, however, we have learned a little too much about our neighbor worlds; it is unlikely that anything like our life will be found on them. (The possibility still isn't entirely zero for Callisto and Titan, perhaps, or even the upper gas layers of the atmospheres of Venus and the gas giants.)

That leaves us the possibility of life on planets circling stars other than our Sun, and those are represented in *Monsters,* too.

To my way of thinking, science-fiction monsters are more frightening than fantasy ones. A dragon that we know doesn't exist isn't nearly as horrifying as the "black destroyer" that someday some explorers of the far distances might really encounter. In fact, a new subdivision of fantasy has arisen in which dragons and other such monsters are so presented as to gain our sympathy. I don't see that happening to science-fiction monsters—at least not as easily.

# 90

# The Power of Evil

The young people living in the United States or some other developed and industrial nation are used to inhabiting a universe ruled by the laws of science.

We know how to control the environment to what we think is our own benefit—to grow food more efficiently, to produce energy, and to control disaster. We know how to prevent many diseases from striking us, how to control or cure them if they do strike. We know how to lower the danger of lightning and how to make planes, cars, and machinery of all kinds quite safe to use.

Even when disaster does strike—when a plane crashes or a tornado hits or someone is murdered or gets an incurable disease—we know there are natural causes and, if we can, we try to find out exactly what those causes are and how to protect ourselves more efficiently against such unpleasant events.

How different things were in prescientific times—and how different they still are in many undeveloped regions today.

When science and modern thought did not exist, and where they do not exist today, the universe is a strange and very frightening thing. There is no knowledge of the scientific laws that govern events. Things therefore seem to take place without natural cause.

Floods come or droughts wither the landscape—storms batter people or epidemics cut them down—lightning strikes or animals die of disease—somehow things go wrong with our most precious plans . . .

Why? Why?

No one in a nonscientific environment even dreams of seeking a natural cause. If something bad happens, it must be because some intelligent being has caused it out of anger or spite. If the event is something no normal human being can bring about, it must be some superhuman being who does it. One of the gods is angry because he or she hasn't been sacrificed to. A passing demon with a hatred for the human race inflicted them. An indifferent spirit is just amusing himself the way a child might when pulling wings off flies. Or perhaps the disaster is brought about by a just and kindly god who has been angered by sin and who wishes to chastise the sinners.

But you don't know, you can't know, exactly what caused the event or how to prevent it. Does one beg the superhuman being for forgiveness, threaten him or her, or make use of certain magical charms or rituals, or what?

And, of course, there is always the suspicion that some people are better informed than others concerning how to handle the supernatural gods and demons. Some people may have learned just how to perform the rituals or how to say the charms in just the right way so as to prevent the supernaturally caused disasters or to bring them to an end.

If these gifted ones are kindly, concerned with the good of the people, they are call priests, seers, saints, or wise men. But what if they are themselves selfish or evil and want to use their control over the supernatural to make themselves powerful or to punish anyone who offends them? Then they are called wizards, witches, enchanters, or necromancers.

Think how dangerous a universe would be if anyone you chanced to meet might, unknown to you, be an enchanter. Some casual thing you say might annoy him, and he might change you into a frog.

Then, too, once you become afraid of any stranger because he (or she) might be an enchanter, it doesn't take much to fear him because he might be a human being with horrifying abilities or habits—someone who looks like a human being but who is so different in various ways that he or she might be considered a "monster."

What if he (or she) is not really alive but a ghost or spirit, an insubstantial remnant of a human being who can take on the appearance of reality but who can disappear at will, and who means evil against you? Or what if he has the ability to change into a wolf (or some other animal) whenever he wants to; or what if he *must* undergo such a change even against his will at the time of the full moon? He is then a "werewolf." What if he eats dead bodies (he is then a "ghoul") or drinks blood (he is then a "vampire")? And what if he lives forever, as long as he can indulge in these appetites or what if he has superhuman strength or other abilities in addition?

In a world in which the idea of scientific law is absent, you don't ask how human cells can change into wolf cells, or how hair can suddenly grow when a man becomes a wolf, and what happens to it when the wolf changes back to a man. You don't think that a diet of corpses might result in food-poisoning, or that an exclusive diet of blood might result in vitamin deficiency or in an iron over-supply.

*Anything* is possible, and as people tell these stories and pass them along, they get more and more horrible and horrifying.

In *Young Monsters* we have collected over a dozen well-done tales about young monsters, those who are children or teenagers. Some are sympathetically, even humorously, told, and some are grisly.

But why should we be interested in such tales? Surely, with our scientific view of the universe, we don't believe that such things as vampires and ghouls and werewolves exist.

Yes, but we can pretend. In fact, that's what makes it fun. In the days when we thought monsters *really* existed, tales of them would have scared us so badly we would have nightmares, or be afraid to go out-of-doors. We would jump at every sound or shrink at every unexpected movement. Such stories would be no fun.

Nowadays, though, we can experience the odd world of nonscience and even get tense or scared *while reading;* but when the story is over, we can dismiss it and return to our normal world, where things happen out of natural cause and where we know what is impossible and what is not. We have the fun of *temporary* fear.

Furthermore, to read monster stories is to move into a world so different from ours that it provides a kind of relief. Our own world has its terrors, too, though they are different from those of the nonscientific world. We don't expect a stranger to be a dangerous enchanter, but we might well fear he is a dangerous mugger. We don't expect to meet a ghost or ghouls when we are passing a cemetary at night, but we might meet a car with a drunken driver at the wheel. We might not expect an angry god or demon to destroy the world in a fit of anger or malevolence, but human beings in charge of governments might destroy the world by nuclear warfare in a fit of fear or anger—or simply misunderstanding.

In a way, it is a relief to turn from the very real power of evil that surrounds us today to the totally different kind of evil that existed in the nonscientific world of ghosts and spirits and enchanters and monsters, as in this collection, *Young Monsters,* which I helped edit.

After all, we know that monsters *don't* exist—and that criminals and war *do* exist.

# 91
# The Devil

Nature is indifferent to human concerns. It is erratic, too. Storms, floods, droughts, blizzards, searing heat, pestilence, and so on may afflict us. And in between these natural disasters, there can equally well be periods of salubrious weather, peace, and happiness.

You might say that so many complex factors must be understood in order to be able to predict these events and put them in orderly perspective that it might be better simply to dismiss natural events as part of the randomness of an impersonal Universe.

That is a hard thing to think, however. In ancient times, when it was especially difficult to forecast natural events, it seemed much easier to suppose that all manifestations of nature were in the charge of invisible superhuman entities (let us call them "gods"), who just happened to be childishly erratic and unpredictable.

A good deal of primitive effort, therefore, was put into determining ways of cajoling gods into being kind. No one ever showed that this cajolery ever did any good in the sense that disasters grew fewer or less intense, but in continued anyway.

Even the logical and rational Greeks, who set up a group of extremely attractive gods, and who viewed them as essentially beneficient, nevertheless imagined that any one of them might suddenly devastate a region if he or she were slighted in some trivial way.

It is, however, wearisome to live with childish gods. It is much more heartwarming to think of them as loving, caring entities who will protect and foster humanity. In that case, though, how does one explain evil and misfortune? The search was out for a scapegoat. Perhaps there was one malignant god who disliked humanity and who was responsible for evil. In the Norse pantheon, there was Loki, the god of fire. Fire itself was a great boon to humanity—but it could also do great harm, and the fire-god partook of this double character. Loki was by far the cleverest of the Norse gods (a pretty dumb lot, particularly as described by Wagner), but he was also a mischief-maker and destroyer.

It was the ancient Persians, however, who developed a particularly dramatic view of the problem of good and evil. In their Zoroastrian religion they pictured the existence of a principle of good, Ahura Mazda, and one of evil, Ahriman. These two were engaged in a cosmic battle, and so evenly balanced were the forces that the addition of the minuscule power of humanity committed to either side would assure the victory of that side. Consequently, humanity participated in this battle, willy-nilly, and was constantly being drawn to one side or the other.

Judea was under the Persian Empire from the sixth to the fourth century B.C., and during that period Judaism picked up this dualistic notion of good and evil. In those portions of the Bible that date back to the pre-Persian period, God is supreme, the author of both good and evil—the evil being either a punishment of sin or a clever device (beyond human understanding) to bring about good for humanity.

In the later portions of the Bible, the idea of evil incarnate is introduced. The Jews called him "Satan," meaning "adversary," since when human beings were judged by God, Satan testified against them as their adversary (or, as we might say, a prosecuting attorney calling for damnation). To the Greeks, Satan became *diabolos,* meaning "slanderer," since he spoke evil even of good men when these stood before the Judgement Seat. And *diabolos* has become our "devil."

Of course, the Jews insisted on the supremacy of God. They could not allow Satan to be equal to God in power, Persian-fashion, and to have the outcome of the battle uncertain. Satan had to be subordinate to God, to have been created by him, and to be controlled by him.

In the Book of Job, for instance, Satan speaks ill of the saintly Job, and God grants him permission to afflict Job, as a way of testing Job's love of God. In the process, Satan not only destroys Job's property, but even his family (killing his numerous sons and daughters) and his health. (I have always thought God is make to look bad in this book. Satan is merely following orders and fulfilling his God-given function, but it is God who allows the destruction of innocent children merely as a way of showing how loyal a devoted follower is. This behavior would never be praised in an ordinary sinful human being, and I don't see how it can be praised in a supposedly all-perfect beneficent God.)

In any case, even though God and Satan are fighting an unequal battle in the Judeo-Christian view, and God's victory is certain and inevitable, the human involvement of the Persian view is kept. Satan is pictured as constantly trying to win over human beings and to possess their souls, even though in the long run such dubious conquests are surely meaningless and won't help him in the least.

So eager is Satan for human souls that he stoops to bribe people, offering them all sorts of worldly advantages in return for their souls. The classic tale of this sort is that of Dr. Faust, which reaches its most sublime form in Goethe's great epic drama.

And yet this has always struck me as silly. It is generally agreed in the Christian view that all men are sinners and that salvation is hard to attain. In the end, it is usually suggested there will be a mere handful of saints going to heaven, while everyone else howls in hell. Why, then, is Satan so eager? He's getting almost everyone anyway. And those few whom God has determined to withold from him are surely beyond his reach.

Then, too, why give these people any bribe at all? The minute a human being consents to deal with Satan and to barter his soul for worldly benefit, he has surely committed a mortal sin. There is no need actually to sign any document with blood or anything else; Satan *has* him (or her) and it's all over.

When I argue in this way, however, I am merely bringing logic to bear on something that is thoroughly emotional.

Human beings *want* the wordly benefits that are traditionally associated with evil. (After all, to grow wealthy—judging from the wealthy of the world—it is necessary to cheat, and scheme, and steal, and slay, and, in short, be thoroughly evil.) However, human beings are also reluctant to pay the penalty that religion calls for in such a case. They don't want to be the beggar Lazarus, and dine in Heaven. They would rather be the rich man, Dives, but *not* burn in hell.

Therefore, writers never tire of describing how human beings can outwit the devil, get benefits out of him, and then refuse payment—and readers never tire of reading about such cases. Even Martin H. Greenburg, Charles G. Waugh, and I never tire of it, so we have prepared a whole book of stories about the Devil, *Devil Tales.* Some, but not all, deal with those devilish documents that human beings are talked into signing (as in Benet's classic "The Devil and Daniel Webster").

Still, I'm sorry for the Devil. Cheating is wrong, even when the Devil is the victim.

# Part VI

# Science Fiction

---

I have had published two collections of essays dealing with science fiction primarily. (After all, I have been deeply involved with science fiction ever since the age of nine.) One is *Asimov on Science Fiction* (Doubleday, 1981) and the other is *Asimov's Galaxy* (Doubleday, 1988). Nevertheless, there are six essays dealing with science fiction that I would like to include here.

The last of these, "Back Through Time," is a review of a motion picture. Ordinarily, I don't do reviews of science fiction, because I don't feel I have the talent to judge someone else's work. In this case, however, I am not considering the theatrical or literary value of the film, only the logic of the thesis—and that, perhaps, I may be allowed.

# 92

# Science Fiction Finds Its Voice

There is no easy agreement as to when we might date the beginnings of science fiction. There are those ambitious souls who lay claim to Plato's tale of Atlantis (c. 350 B.C.) as the origin, and others, even more all-embracing, who wish to annex the *Epic of Gilgamesh* (c. 2800 B.C.).

This is, however, at least in my opinion, simply nonsense. Science fiction must involve itself with science and technology at least tangentially. It must deal with a society noticeably different from the real one of its times, and this difference must involve some change in the level of science and technology. If this is so, science fiction cannot predate popular awareness of the connection between advancing science and technology and social change, and that brings us up to the Industrial Revolution. Anything earlier is only *fantasy,* even if it involves trips to the Moon, as does the *True History* (c. 150) of Lucian of Samosata.

There are those who accept this view and consider that science fiction dates back to the early decades of the nineteenth century. Some suggest Mary Shelley's *Frankenstein* (1818) as a beginning. Others, however, feel that *Frankenstein* is more honestly classified as a "Gothic tale" in the tradition of Horace Walpole's *The Castle of Otranto* (1765). Later examples of what are often thought of as early science fiction might be so labeled as well, notably the works of Poe and Hawthorne.

It might seem, then, that we ought to start true science fiction with Jules Verne's *Five Weeks in a Balloon* (1863). Verne wrote science fiction without Gothic trappings, and he was the first person to write science fiction almost exclusively, and to gain great popularity and wealth as a result. Certainly, 1863 as a beginning sounds good.

Yet we can present an argument against even this. Science fiction, whether it begins in 2400 B.C., A.D. 150, 1818, or 1863, has always been a thin thread through literature generally. Relatively few authors have tried their hand at anything that can be called science fiction by even a liberal interpretation of the term, and even fewer have done so in truly popular fashion—Jules Verne and H. G. Wells are the two best we can name prior to the 1920s.

Why not then find a beginning to science fiction as a mass phenomenon? When did science fiction begin to be turned out in quantity, first by dozens, then scores, and then hundreds of writers? What set it on the road to where it is today, an extraordinarily popular literary phenomenon that has many first-rate luminaries? One need only mention Robert Heinlein, Arthur C. Clarke, Anne McCaffrey, Frank Herbert, Ray Bradbury, Ursula K. LeGuin—but modesty forbids my continuing.

Clearly, what is responsible for this is *magazine* science fiction, which began with the first issue (April 1926) of *Amazing Stories,* published by Hugo Gernsback.

There are those who will object strenuously to the idea that magazines were crucial to the development of the genre of science fiction. Writers and critics who demand literary "respectability" for science fiction consider magazine science fiction to have been a "ghetto" that reduced the genre to a variety of pulp fiction (magazines that published lurid stuff for semi-literates were often printed on cheap, pulpwood paper), so befouling the genre as to cause mainstream writers to refuse to deal with the field.

There is something to this. Certainly, some 90 percent of magazine science fiction was indeed teenage childishness at first. But remember Sturgeon's Law: 90 percent of *everything* is crud. Nevertheless, magazines did create a forcing ground in which many youngsters sharpened their talents who otherwise would never have entered writing at all, or who would have written something other than science fiction. "Literary" science fiction *never* made the field popular, although it made a very few writers popular. It was magazine science fiction that did the job, although it did have to crawl before it could walk, and walk before it could run.

To turn up one's nose at science-fiction magazines, then, does the phenomenon a great injustice and merely reveals the nose-up-turners to be pretentious pedants.

Let us, then, turn to magazine science fiction. It did not have an easy birth. It was precisely because the field was not popular and had few practitioners that there were not enough writers to support a monthly magazine. Gernsback had to depend on reprints of H. G. Wells and Jules Verne in his early issues, and only gradually did new writers rally to the field.

These new writers were sometimes raw beginners whose skills were, as yet, small. Other new writers were pulp writers who turned from their adventure yarns (or whatever) to try their hand at a new variety of pulp without much understanding of what science fiction really was. Consequently, the magazine science fiction of the twenties does not offer as much in the way of quality.

At that time, as in all times before or since, the longer the story, the better the quality on the whole (but only on the whole, for there were many notable exceptions, of course). It is not surprising, then, that the one notable example of magazine science fiction of the twenties was *The Skylark of Space,* a novel serialized in *Amazing Stories* in 1928.

Unfortunately, because we cannot include novels in this collection, we must confine ourselves to "novellas," which are stories that are rather long but are short enough to be included in a single issue of a magazine. Since more novellas than novels can be published in the magazines simply out of consideration of length, some of the best writers in the field began to concentrate on novellas—as you will see when you read *Best Short Science Fiction of the Decade: The Thirties.*

It was in the 1930s that magazine science fiction began to find its voice. Even where the writing remained "pulpish"—that is, overwritten and unsubtle—ideas began to flourish, and the minds of the readers were stretched.

In the thirties, the leading science-fiction magazine was *Astounding Stories.* It had begun with the January 1930 issue and had quickly outpaced *Amazing Stories,* both because it offered higher word-rates and because its editor, Harry Bates, abandoned Gernsback's didacticism and emphasized action. With the March 1933 issue, however, at the depth of the Great Depression, William Clayton, the publisher, went bankrupt. The title was bought by Street & Smith Publications, and the magazine was resurrected with the October 1933 issue under the editorial direction of F. Orlin Tremaine.

Tremaine remained at the helm for four years, and his great contribution was the notion of "thought-variant" stories—that is, stories featuring some startling new notion, or some unexpected variation of an old notion. These greatly pleased the readers, and the magazine was never again threatened by failure.

As an outstanding example of a thought-variant, there is "Sidewise in Time," by Murray Leinster (a pseudonym of William F. Jenkins), which was published in the June 1934 issue of *Astounding Stories.* "Sidewise in Time" was the first attempt to deal with the notion of parallel time-streams, of universes that, at key moments, could take one path or another, with all paths attaining some sort of existence. (Forty years later, physicists dealing with the more esoteric consequences of quantum mechanics are finding themselves forced to deal with this Leinsterian concept.)

Another unusual story from the Tremaine era is "Alas, All Thinking," written by Tremaine's editorial predecessor, Harry Bates. It appeared in the June 1935 issue of *Astounding Stories.* Bates did not write many stories, but the few he did write were good. "Alas, All Thinking" is a chilling tale of evolutionary degeneration.

While Tremaine's *Astounding* utterly dominated the field, occasional good stories appeared in the other magazines of the time, like *Amazing Stories* and *Wonder Stories.* One outstanding example among the longer stories is "He Who Shrank," by Henry Hasse, which appeared in the August 1936 issue of *Amazing Stories.* Hasse's story was a true thought-variant and could easily have graced the pages of Tremaine's magazine. As the title indicates, it deals primarily with a shrinking human being, something that has been explored both before and after this story—but Hasse, as you will see, has an added concept entirely peculiar to him. Oddly enough, Hasse wrote very little, and this is the only story of his to achieve any fame.

The most outstanding writer of the Tremaine era was John W. Campbell, Jr. He began his career as a writer of "super-science" stories, in imitation of E. E. Smith, and was second only to Smith in that field. But then, under the pseudonym of Don A. Stuart, he took to writing much more subtle stories, with surprisingly high literary and emotional content. His first of this type was the short story "Twilight," which appeared in the November 1934 *Astounding Stories.*

The longest and best Stuart story, however, was "Who Goes There?", which appeared in the August 1938 issue of *Astounding Stories.* It is included here—and if you have never read it before, I envy you, for you'll find it among the cleverest and most insidiously horrifying stories you have ever read. I won't say a word about the plot. You must find that out for yourself.

By the time that "Who Goes There?" appeared, however, another revolution had taken place. In December 1937 Tremaine was promoted to a higher position, and John W. Campbell himself became editor of *Astounding Stories.* He quickly changed the name of the magazine to *Astounding Science Fiction* and began to search for writers capable of achieving greater heights of literacy and greater fidelity in their description of science and scientists.

To begin with, of course, he had to use, and encourage the further development of, authors already in the field. Horace Gold had written several good stories for Tremaine under the pseudonym of Clyde Crane Campbell. Obviously, he couldn't use that pseudonym under the new dispensation. In the December 1938 issue of *Astounding Science Fiction,* then, Gold's first story under his own name appeared, "A Matter of Form," which offers an astonishingly realistic description of the adventures and misadventures of a man who was hampered (as the title indicates) by his form.

Another Tremaine author who was ripe for greatness was L. Sprague de Camp. His first published story had appeared in the September 1937 *Astounding Stories,* and he has been writing numerous stories since. He knew a lot about science and history and was meticulous in his accuracy in both fields. Furthermore, he was one of the relatively few science-fiction writers with a good sense of humor. He came into his own when Campbell introduced a sister magazine to *Astounding Science Fiction—Unknown,* which first appeared in March, 1939. It featured "adult fantasy," as meticulous in its internal logic as the stories in *Astounding Science Fiction.*

De Camp quickly became a mainstay of the magazine, and his amusing "Divide and Rule," with its popular but logical blend of medieval chivalry and modern technology, appeared in its second issue, April 1939.

Thus, we present half a dozen of the best stories of the thirties, and we hope eventually to return with an (at least) equally good selection of stories of the forties.

# 93

# The Five Greats of Science Fiction

Fantasy, which is sometimes confused with science fiction, has a long and resplendent history, for all fiction is, in a way, fantasy. We can trace great literary fantasy to Homer's *Odyssey,* and beyond that to the legends of the early chapters of the Bible, and beyond that to the Sumerian tale of Gilgamesh, which dates back to about 2800 B.C.

Science fiction is that branch of fantasy that bases its plot convolutions on changes in the level of science and technology, and it could not really be written until the world came to realize that scientific and technological advances were changing society—that is, not until about 1800, when the Industrial Revolution was well under way.

Some think that the first true science fiction novel was Mary Shelley's *Frankenstein,* which was first published in 1818, when the author was only twenty-one. Though it reads very much like juvenilia, it makes use of science, rather than magic, to create artificial life, and that is the crucial point.

She was followed by Edgar Allan Poe, who wrote occasional pieces of science fiction but who is best remembered today for his horror tales.

The real founder of science fiction, and the first of my five "Greats" in the field, is, however, the Frenchman Jules Verne. He was the first to devote himself almost entirely to science fiction, the first to support himself well while doing so, and the first to bring science fiction to the attention of a vast public.

Verne was the son of a lawyer, and he rebelled against his father by deliberately adopting a radically different lifestyle. He tried to run away to sea, and always remained fascinated with ships. He tried to become a conventional writer and was a prolific failure. He became a stockbroker and hated it.

Finally, in 1863, when he was thirty-five years old, he was inspired by his reading of Poe to write an adventure story entitled "Five Weeks in a Balloon," which, much to his own surprise, proved an enormous success. It was essentially a travel story, but it dealt with an unusual mode of travel. Verne at once capitalized on his success by writing other such novels, which he called "Extraordinary Voyages," taking his characters to the center of the Earth, to the bottom of the sea by submarine, to the Moon, to the far reaches of the Solar System by hitching a ride on a comet, and so on. His greatest success was less extraordinary than any of these cases, for it dealt merely with the circumnavigation of the world: *Around the World in Eighty Days.*

Verne attempted to deal meticulously with the scientific background of his stories. Even though he was sometimes wrong (as in thinking it was possible to shoot a vehicle into space by way of a giant cannon without instantly killing the people on board), he imbued his books with a great air of reality.

With Verne, science fiction became a recognized branch of literature. His popularity spread throughout the Western world, so that my father read him in Russian and thirty years later and five thousand miles westward I read him in English. We read him with equal avidity, and it was only in his works that our science-fiction tastes met.

Verne, however, kept one foot (and sometimes both) firmly on the ground at all times. He made use of no marvels and did his best to cling as much as possible to the already known.

If science fiction was to advance, it would have to be freed from these chains; its imagination would have to soar. That brings us to the second of our five Greats: the English writer Herbert George Wells.

Wells, like Verne, was born a member of the middle class, but rather lower down in the scale. Wells was the son of a shopkeeper who had married a servant. Again like Verne, Wells's early life was rather a litany of failure. He resented his lowly status in the British social caste system of Victorian days, so he became strongly socialist. He had a sketchy school career, which he filled in by an enormous program of self-education. He was not in good health, and had an unhappy marriage (his sex-life, however, was hectic and unquiet throughout his life).

Finally, inspired by Verne, as Verne had been inspired by Poe, Wells wrote a book of a new kind in 1895, when he was twenty-nine years old. It was *The Time Machine.* It made a tremendous hit, to Wells's surprise. In a way, he seemed to be following in the steps of Verne, for *The Time Machine* was a travel tale; but there was a difference. It was a trip into the far future. It did not make use of known technology or of contemporary technology with minimal improvements. It flung itself into the unknown by describing something that was utterly unlike anything that existed. Wells had written the very first time-travel tale. (It was not a matter of Mark Twain's Connecticut Yankee being driven into the past by a knock on the head. Wells's hero moved at will along the time dimension by machine—as though he were driving a locomotive back and forth along its rails.)

Other books and short stories followed. *The Island of Dr. Moreau,* published in 1896, was a tale of biological change. *The Invisible Man* (1897) was one of his most remarkable short stories. Then, in 1898, came his most influential book, *The War of the Worlds*—the first story of extraterrestrial invasion and interplanetary warfare. And when he wrote his tale of a trip to the Moon (as all early science-fiction writers did), he used neither rockets nor cannons, but made use of a gravity-shield.

Wells's use of free-wheeling ideas quickly made him more popular than the aging Verne, who reacted with bitterness. Verne pointed out that his cannons existed, but Wells's gravity-shield did not. But, of course, that was exactly the point that worked in Wells's favor. After 1900, Wells turned increasingly to successful mainstream fiction, and his *Outline of History,* published in 1920, was a super bestseller. And yet now, nearly a century after *The Invisible Man,* it is his science fiction for which he is best known and most ardently remembered. He may very likely have been the most influential science-fiction writer who ever lived.

As the twentieth century wore on, it seemed that no one would arise to replace Verne and Wells, but in 1926 a new phenomenon appeared on the scene—a magazine, *Amazing Stories,* devoted solely to science fiction. Until then, science fiction was the province of the occasional writer, but now there was an outlet that allowed beginners to try their wings (virtually without payment, to be sure).

There were no immediate results, of course. In the first years of its existence, *Amazing Stories* had to fill most of its pages with reprints by Poe, Verne, Wells, and some lesser lights. And then came the third of the science-fiction Greats, Edward Elmer Smith.

Smith had written a novel called *The Skylark of Space* around the time he was getting his Ph.D. in food chemistry (so that when his stories finally appeared, it was always under the name "E. E. Smith, Ph.D."), but there was no place where it could appear, for the writing was unskillful, the plot was wild, and, to begin with, there was no market for such things.

*Amazing Stories,* however, was meant for just this, and the market met its man when *The Skylark of Space* appeared in its pages in 1928. Smith was thirty-eight years old at that time. The readers exploded with joy. And magazine science fiction, having its first "superstar," was no longer a possibly passing phenomenon. It was here to stay.

Smith's story was the first ever to deal with *interstellar* flight. It had "superscience," "superweapons," and "superheroes." It was a "wild west story" wtih spaceships and blasters. It had a rapid-fire, expanding plot that established the tradition of the "space opera" (by analogy with the "horse opera," as Westerns were called).

Smith continued to write his space operas, steadily increasing the scope and wildness of action, and he was imitated by others, until the whole sub-genre began to sag and crumble under its own weight. Nevertheless, Smith had established an air of optimism in science fiction, a feeling that human beings could face all the vastness of the Universe unafraid, bringing all of it under its domination. This feeling was to particularly characterize American science fiction for a generation.

In a sense, Smith carried the excitement of Wells to the ultimate extreme, but it was at the price of abandoning any sense of realism, something to which Smith's imperfect prose contributed.

The scene was set, then, for the fourth of the Greats of science fiction, Stanley Grauman Weinbaum, who was ten years younger than Smith. He was a chemical engineer, who, like Verne, Wells, and Smith, turned to writing early in life and was, at first, unsuccessful.

After 1930, *Amazing Stories* was no longer alone in the field. Other magazines entered the field that Smith had set to growing, and in 1934 Weinbaum sold "A Martian Odyssey" to *Wonder Stories,* the least of the three magazines then being published. He was thirty-four years old at the time.

"A Martian Odyssey" was on a much smaller scale than anything Smith or his imitators wrote; it was merely the report of the first human expedition to Mars. In comparison to the space opera, however, the writing was clear and realistic, and the fustian was gone.

What is more, Weinbaum labored to describe Mars in sensible terms. Until then, extraterrestrial creatures had usually been superintelligent villains, or superdreadful beasts, always enormous menaces that had to be enormously killed. Weinbaum produced an engaging Martian who was as intelligent as a human being, but not intelligent *in the same way* as a human being. This creature virtually stole the show. Weinbaum described other Martian life-forms, too, each fascinating and sense-making in its own way.

Weinbaum was, in my opinion, the best science fiction writer since Wells, and the readers responded to his style with delight. He rapidly wrote other stories, which were accepted by *Astounding Stories,* now the leading magazine in the field, and for a year and a half it was generally acknowledged that he was the most popular writer (as well as the best) in the field.

But only for a year and a half. Toward the end of 1935, Weinbaum, not yet thirty-six, died of cancer. Where he might have taken the field, we will never know.

But now there came the fifth Great of science fiction. In a way, he meant more than all the rest. Till then, the Greats had arrived unpredictably and worked singly, but it was John Wood Campbell, Jr., who put an end to that.

Ten years younger than Weinbaum, Campbell was trained in physics at M.I.T. and Duke. Unlike the other Greats, he was successful in his writing from the start. His first published story, "When the Atoms Failed," was published in *Amazing Stories* in 1930, when he was only twenty years old. In the next few years, he quickly made his reputation as writer of space-opera second only to Smith. In 1934, with the publication of his story "Twilight"—under the pseudonym of Don A. Stuart—he began a second reputation as a writer of more subtle, emotional stories.

His true career, however, began in 1938, when, at the age of twenty-eight, he became editor of *Astounding Stories,* whose name he quickly changed to *Astounding Science Fiction.*

As editor of the world's leading magazine in the field, he had a great deal of power—and he began to use it, inexorably, to refashion science fiction in the direction he thought it ought to go. He didn't want mere adventure. He didn't want space opera (though he continued to publish Smith). He wanted stories that dealt with thoughtful extrapolations of science and engineering. He wanted characters who, if they were scientists and engineers, acted and thought as though they were scientists and engineers. He pushed for what has come to be called "hard science fiction," in which the author does not play fast and loose with scientific principles.

In short, where Verne established the field, where Wells added ideas, where Smith added wild optimism, where Weinbaum added reason, Campbell added scientific respectability.

Campbell did more. With infinite pains, he cultivated new authors in whom he saw promise, giving them ideas, analyzing their efforts, encouraging them to try again, until he had gathered about himself an amazing constellation of skillful young writers who dominated the field for a generation. Three of them have, in fact, been active for almost half a century, and they are *still* familiarly known as "the Big Three": Robert Heinlein (who died in 1988), Arthur C. Clarke, and yours truly.

Campbell's success made fiction expand in all directions once World War II was over. The number of writers multiplied; the number of magazines multiplied. Science fiction began to appear in hardcovers and softcovers put out by major publishers, and then it appeared on the bestseller lists. Science fiction began to make blockbuster appearances in movies and television, too. And the greatest of all wonders (to someone like myself, who began half a century ago) was this: Science fiction writers began to make fortunes out of their writing.

Since Campbell the field has become too large to be dominated by anyone. There are dozens of "greats" now, but no "Great." That is the measure of what the five Greats have done for the field.

# 94

# The Success Of Science Fiction

Science fiction has had many faces, and with every one of them it has been, in one way or another, successful.

We can reach very far back, if we are willing to broaden the definition of science fiction to have it include any tale that was fanciful and imaginative in terms of what was known of the Universe in its own time. Consider the audiences of a time when writing was unknown or, at the very least, restricted to a very few. If so, we can easily picture a bard striking chords on his lyre as he chants the tales of Odysseus's wanderings to an illiterate, and raptly attentive, audience. Nearly two thousand years earlier still, a bard would have sung the feats of Gilgamesh.

Successful? Can we doubt they held their audiences enthralled when, in written form, these tales survive, and are admired, to this day?

What of later times, when writing had become part of the intellectual armory of all educated people? Then science fiction, in common with all literature, had as its chief, and almost only, form that of the written tale. If we continue to think widely enough to have science fiction include fantasy, then from days of Vergil and Ovid, down through the knightly legends of the Middle Ages and the ghostly Gothic tales of early modern times, we have a kind of "proto-science fiction" that was invariably among the most popular and successful forms of literature of its time.

By the beginning of the nineteenth century, the Industrial Revolution was in its early stages, and it became clear to thoughtful observers that the world was changing rapidly and that, powering that change, was the force of advancing science and technology. Science fiction proper came into existence: stories that dealt not only with the fantastic, but with the fantastic that *might conceivably be made possible by further scientific advance.*

At first, little of this new and significant variety of fantasy was written. Perhaps the first of the type was Mary Shelley's *Frankenstein,* published in 1818, which is still popular now.

For half a century, occasional science fiction was written by important writers, notably Edgar Allan Poe. Then, in 1863, the Frenchman Jules Verne published the first of what he called his "Extraordinary Voyages," one that was entitled *Five Weeks in a Balloon.* He poured out a stream of other books of this sort, and his fame rang through the world for the remainder of the century. He was the first person to write science fiction primarily, and the first writer to make himself both famous and rich by such writings.

What's more, he lived to see the Englishman H. G. Wells outdo him. Wells, a more gifted writer, with an exuberance of ideas, made himself more famous still.

By Wells's time, however, literacy, at least in Western Europe and in the United States, had become the common property of most of the population. With mass literacy, there came the phenomenon of mass literature—cheap books and magazines, hastily and crudely written perhaps, but intended to satisfy the needs of the vast numbers whose education and sophistication were limited.

Eventually, the Luxemburg-born publisher, Hugo Gernsback, took what now seems an obvious step. He had been publishing occasional science fiction that dealt with radio and with other technological marvels of the new century. In 1926 he produced *Amazing Stories,* the first magazine in the history of the world to be devoted to science fiction and nothing else. (Sixty years have passed and the magazine still exists.)

At first, the only stories Gernsback could include in the magazine's pages were reprints of Poe, Verne, Wells, and a number of lesser lights. Little by little, however, writers for other magazines of the era tried their hand at the new form. Enthralled readers began to try to write, too.

Slowly, "magazine science fiction" grew in popularity. It was a new kind of science fiction, rather crude and rough at first, rich in excitement and stereotypes, low in characterization and subtlety—but young people loved it. I came across my first science-fiction magazine at the age of nine and was lost forever.

The initial audience grew older, a change that was accelerated by the fact that a still newer face of science fiction arose, which drew off the youngest of the readers. It was just as easy to print graphics as words, and the new century saw the beginning of the "comic strip" in newspapers. It was inevitable that some of these would be science fictional. Such strips as "Buck Rogers" and "Flash Gordon" had more readers during the 1930s than the magazines did.

Comics made their way into magazine form. At first they were reprints from the magazines, but inevitably some began to publish original strips. In *Action Comics,* the strip "Superman" appeared, and soon it became far and away the most popular science-fiction item that had ever appeared in that form.

The visual side of science fiction was not confined to the motionless world of the comic strip. Movies existed, and, almost from the start, attempts were made in the direction of science fiction. The first science fiction movies, by the very fact of the complexity of the medium and the imperfect way in which it could be controlled, were as crude as the first examples of original magazine science fiction. The early "Flash Gordon" serials, for instance, were little more than photographed comic strips.

There were successful attempts, however, to produce science-fiction movies of surprising quality for their times. Thus, we have *Metropolis,* shot in Germany in 1926 (the year that saw magazine science fiction come into existence) and directed by Fritz Lang. Then, in 1936, came the curiously prophetic British film *Things to Come,* made from an H. G. Wells book.

Magazine science fiction was improving rapidly in quality. In 1938 John W. Campbell, Jr. became editor of *Astounding Science Fiction* and strove, successfully, to improve the scientific and technological backgrounds of the stories in the magazine. He gathered about him a new group of young writers, most of whom had scientific backgrounds (including me).

The use of the nuclear bomb at the end of World War II shocked the whole world into a new attitude concerning science fiction. It was no longer "kid stuff." The new breed of science-fiction writers had forseen the bomb, and during the war they had written many stories dealing with it. As a result, the field gained a new respect. The development of rocketry during the war and the beginning of talk about ventures out into space deepened this respect.

The improved reputation of science fiction made itself felt in the movies, of course. In 1950 *Destination Moon* was made. Part of its importance lay in the fact that it was made from a book by Robert Heinlein, the most important of John Campbell's stable of writers, and that Heinlein himself acted as adviser on the film.

Although most science-fiction films in the early post-war era were "monster" films, with hordes of Godzillas and giant spiders filling the screen, *Destination Moon* attempted to portray a rocket flight to the Moon realistically.

While the film was reasonably successful when it came out, *Destination Moon* has not stood the test of time. For one thing, the special effects are primitive. In fact, they have been the limiting factor in *all* visual science fiction. In print, you can have whole stars explode with the flourish of a few sentences; you can describe strange life-forms and such phenomena as weightlessness. *Showing* them is a lot harder.

In fact, radio was, at first, the most successful nonprint science-fiction medium, for like print it did not have to *show* anything. Sound effects were enough. The old *Buck Rogers* programs on radio are still remembered affectionately by those who heard them in the 1930s.

However, the technology of special effects was advancing. In 1966, *Fantastic Voyage* appeared, with splendid simulations of the interior of the human body. Then, in 1968 came *2001: A Space Odyssey*. It was more than a bit slow, and its ending was as obscure as any ending has ever been. Nevertheless, its simulations of space flight and the feeling it gave of a giant computer were marvellous. It became the first true hit in the history of movie science fiction.

Television, with less money to spend on a single show, was far behind the movies in special effects. It did not produce a true science-fiction program of any quality until 1966, when "Star Trek" appeared on the screen. It's popularity caught even its producers by surprise. When an attempt was made to cancel it after the first year, a viewer revolt of astonishing proportions forced its reinstatement. Then, when it *was* brought to an end in 1969, it continued in reruns—as it does to this day—and gave birth to a series of motion pictures.

Meanwhile, science fiction in print profited by the growing familiarity and affection of the public for the field. Beginning in 1949, major publishing houses began to put out science-fiction novels, which were then reprinted by major paperback houses. By the 1970s science-fiction novels were beginning to appear on the bestseller lists, and some science-fiction writers who were thoroughly used to a life of poverty and obscurity were astonished (and not necessarily displeased) at suddenly finding themselves rich and famous.

But then, in 1977, came *Star Wars,* and the astonishing climax of it all. Special effects had become so advanced, so skillfully done, so well handled, that one could finally see what science fiction till then had only been able to describe. It is perfectly possible to argue that special effects drown out the subtleties of the field and put a premium on superficiality, but those special effects are what people want.

*Star Wars* at once became one of the most popular and most profitable moving pictures ever made, and it set off a new fashion in such blockbusters. It was more popular by far than any story in print had been, than any comic strip had been, than anything earlier seen in the visual media had been, and it was to be rivaled only by successors of the same type, like *The Empire Strikes Back, E.T.,* and so on.

This popularity makes possible—indeed, necessary—a book like *The Science Fiction Image,* which gives an encyclopedic overview of science fiction in the nonprint media. I trust that new and, inevitably, enlarged editions will come out in the future at periodic intervals.

# 95

# Science Fiction Today

Science fiction has changed enormously since I began writing it, professionally, over fifty year ago.

When I submitted a story for the first time on June 21, 1938, there were just three magazines in the field, and only one of them, *Astounding Science Fiction,* was any good. Amateur publishers used to put out tiny printings of some poorly written science-fiction novels every once in a while. There were a few comic strips, notably *Buck Rogers* and *Flash Gordon,* along with an occasional very primitive movie serial.

But now?

In the print media, science-fiction novels are commonly found on the bestseller lists, both in hardcover and softcover. The bookstores have shelves full of them. The movies and television find science fiction to be profitable blockbusters. Science-fiction courses are taught in high schools and colleges. Short-story anthologies exist by the hundreds. Science fiction is *big time.*

It might seem to you, then, that it must be a great deal easier to break into the science-fiction field now than it was fifty years ago. After all, the target is so much larger now.

Unfortunately, I don't think that's so. Let us analyze the situation more closely. Fifty years ago, when science fiction (s.f.) consisted of three magazines and virtually nothing else, there were many other outlets for fiction. It was the heydey of the pulp-magazine craze. Every newsstand had dozens of them, in every conceivable category: romance, mystery, western, jungle, war, horror, adventure. Some came out monthly, and some biweekly, and some even weekly. There were also "slick magazines," which published a great deal of fiction and paid much more than the pulps did.

Of them all, the science-fiction magazines were the smallest, the least lucrative, the most specialized, and the least-regarded segment. Almost none of the myriads of young people who had the itch to write considered science fiction as a viable outlet. The science-fiction magazines drew their new prospects from among their own long-time fanatic readers, who had been reading science fiction since they had learned to read and had no interest in anything else. They didn't care for either fame or wealth, only to write that wonderful stuff they were reading and see their name in print in a science-fiction magazine. There weren't many of those fanatics (usually abbreviated as "fans"), but I was one of them. I had been reading science fiction avidly from the age of nine, and I was eighteen when I made my first sale.

Under those circumstances, it was not necessary to be a great writer, you understand. There were few science-fiction writers of any kind in those days, and still fewer good ones. If you were eager to write science fiction, and knew grammar and spelling, and had read enough science fiction to know a new idea from an old one, that was about all that was needed.

Nowadays, all that has changed. In the first place, the fiction market in the print media has contracted violently in the last fifty years (due to the coming of the comic magazine, and then, even more important, television). The pulps are gone. What slicks exist publish very little fiction. In fact, the only branch of popular fiction that has expanded wildly in the last half century has been science fiction. (Mysteries and romances have done no more than hold their own over the long run.)

This means that of all the youngsters who *today* grow up with the itch to be writers, a sizable percentage tend to flood into science fiction. There are hundreds of excellent science-fiction writers today where, half a century ago, there were mere dozens of not-so-excellent ones. In addition, many of those who entered the field years ago are still there. The "big three"—Arthur C. Clarke, Robert A. Heinlein, and myself—whose books are sure-fire bestsellers today, have each been writing steadily for nearly fifty years. Clarke, despite his advanced years, shows no signs of slowing down, and I, of course, am still a youngster. (Robert Heinlein passed away in 1988.)

What's more, all these writers tend to write novels. That's where the money and fame are. And novels are precisely what a beginner would find it difficult to do. A novel possesses a complex structure, with interlocking plots and subplots; characterization is expected to be developed and dialog is expected to show a certain depth and wit. If a beginner throws caution to the winds and determines to tackle a novel anyway, he or she finds it represents an enormous investment of time and effort, all of which (chances are) will be thrown away—except for what good the writing experience will do him or her.

The natural way in which science-fiction writers broke into the field in my early days was to turn out short stories for the magazines. (There was, after all, nothing else to do, since only two or three novels—at most—were published each year as magazine serials.) Clarke, Heinlein, and I all got our start as writers of science-fiction short stories for the magazines. We worked our way up to novels by stages.

Well, then, are there not science-fiction magazines that publish short stories today—to say nothing of science-fiction anthologies?

Yes, but skip the anthologies. The vast majority of them include reprints—stories that have already appeared in the magazines. That leaves only the magazines.

Unfortunately, the magazines have not expanded along with the rest of the field. There are four magazines today that specialize in science fiction. In order of age, they are *Amazing Science Fiction Stories, Analog Science Fact-Science Fiction* (which had once been *Astounding Science Fiction), The Magazine of Fantasy and Science Fiction,* and *Isaac Asimov's Science Fiction Magazine.* In addition, there are a couple of other magazines that publish some science fiction. Most notable of these is *Omni,* which publishes two or three stories in each issue. It pays much higher rates than the others do, and consequently it seeks its stories from among the established writers.

The magazine field, therefore, is not much larger than it was fifty years ago, and the competition is keener. The level of writing in the magazine is therefore substantially higher than it used to be. My eighteen-year-old self, if transported into the present with no more talent than I possessed then, might not have been able to break in.

However, all is not lost. In the old days, when a writer established himself (or herself) as a science-fiction short-story writer, he stayed there (having nowhere else to go), leaving that much less room for newcomers. Nowadays, as soon as a science fiction writer begins to make a name for himself in the magazines and has gained the necessary expertise, he shifts to writing novels. The result is that the magazines are forced to be on the continual look out for new young writers. These new writers have to be good, to be sure—it is no longer enough that they feel warm to the touch—but the fact that they are unknown is not held against them.

But so what if the competition is keen? That makes the task the more challenging, and the triumph sweeter in the end. The rules are the same. You have to read a great deal of science fiction so you can gain some insight into what science fiction is and what makes it good. And you have to write a great deal of science fiction, because only by writing can you gradually learn the tricks of the trade. And you have to have an inhuman perseverance and develop a thick skin against disappointment and frustration. And don't think the world is picking on you. I suspect that Homer and Aeschylus had all the same experiences you had when it came to getting started.

Perhaps something else occurs to you. It may seem to you that when I was just beginning (back in the Middle Ages), hardly anything was known about science and I could write freely about interplanetary travel and robots and all that stuff. Nowadays, however, we *have* interplanetary travel and robots, so what is there to write about? Hasn't science caught up to all the science-fiction plots? Isn't science fiction dead?

Not at all! Nohow! The science-fiction writer is tied to the front end of a locomotive that is speeding across the landscape. No matter how far and how fast the locomotive is going, the writer is looking ahead and sees an endless vista.

Indeed, scientific advance gives us the inspiration for fascinating new backgrounds. We used to think Mars had canals; now we know (not "think") it has extinct volcanoes. And we know Io has active volcanoes. And we know that Venus is as hot as hell—literally—and has no oceans. We can get rid of the tired old planets and make use of brand-new ones, and have the satisfaction of knowing that there's less guesswork and more knowledge now.

Again, think of all the new concepts science has given us. We have neutron stars, and black holes, and quasars, and exploding galaxies, and big bangs. We have mesons, and hyperons, and quarks, and gluons. We have DNA and biogenetics. We have computers and microchips. We have jet planes and satellites of every kind and probes and shuttles. We have seen closeups of Uranus and its satellites. We had *none* of that when I was starting out.

When I think of all the items science has given us to play around with now, and how little I had back in 1938, I stand amazed that I was able to think up any stories at all in that medieval period.

Of course, we have to be careful of fashion. When I first started reading science fiction, it was all adventure and Sunday supplement science. It was written in primary colors and in jagged lightening streaks. It was ideal for a bright nine-year-old to get started on.

By the time I began to submit stories, however, it became fashionable to load them down with authentic science and to try to make the characters sound like real scientists and engineers. The 1940s and 1950s were the heyday of "hard science fiction," and that was my forte—and (to tell you the truth) I still write it, even though it sounds old-fashioned today.

In the 1960s, there came a period of stylistic experimentation called "the new wave," which, it seemed to me, made hard reading and wasn't very successful. However, it settled down into the literary style we have today.

So however much you may want to read the "old classics" (like Asimov), and however much you may enjoy them, too, you had better also read, and pay close attention to, the kind of material that is appearing in the magazines *now*. That is what you should be writing.

Of course, you may be asking yourself if you should be writing for the print media at all. Shouldn't you be breaking into movies and television, where the BIG money is?

Frankly, I don't know how that's done. I've never worked in the visual media myself, except on two or three minor occasions when I was talked into it much against my will.

From what little I know of it, I suspect that while movies and television have the potential for BIG money, that's all they have—potential. They also have the potential for including ulcers and lunacy, and that potential is much more frequently realized. This is just my opinion, you understand, and what I have observed and heard in connection with those of my friends who have been caught up in the visual vortex.

It is my experience that when you write for the print media, what you write is what gets published. If there is a need for revision, the editor asks *you* to revise, and the chances are even good that you will get to see a galleyproof so that you can make sure that any last-minute editorial changes meet with your approval.

What you write for the visual media, however, must apparently meet the requirements of the producer, the director, the various actors, the office boy, strangers who pass in the street, and the mother-in-law of any or all of these, each of whom changes your product at their whim. If you are a real writer, money isn't going to compensate for never being free to write as you wish.

Anyway, just one final word. If you are going to write science fiction short stories, don't forget to try *Isaac Asimov's Science Fiction Magazine*—and take out a subscription to it, so you'll know what we're after. I admit this is a plug, but what the heck.

# 96

# The Feminization Of Science Fiction

There is no question about the boom in science fiction. From *Star Wars* to *E.T.*, science fiction has recently been the hottest thing on the screen.

That, however, is the point. They are on the *screen,* and what has made them screaming successes are the technological advances that have made stunning special effects possible on that screen. What effect, however, has this success had on the older field of *printed* science fiction?

It has not (unfortunately) made instant millionaires out of all of those who write science fiction for magazines and books, because the tens of millions who scramble to the movie houses to see those special effects do not run with equal eagerness to read mere words.

*But some do!*

A small percentage do, and that has made an enormous difference. Let me explain.

Magazine science fiction had its beginnings in 1926 and for a third of a century—a full generation—it was a strongly masculine field. The readers were almost all young men, and, to a large extent, rather *withdrawn* young men, who had either not come to be at ease with members of the opposite sex or who were actively afraid of them.

This is not to say that young women did not read science fiction. A few did. So few, however, that their visibility was low to the point of nonexistence. The result was that the stories that were written demonstrated a purely masculine point of view. If female characters were included at all, they were, stereotypically, total nonenties—physically attractive cyphers—whose role was to get in the way of the hero and to require rescue at inconvenient moments. (Dale Arden in the *Flash Gordon* comic strip is exactly what I mean.) And even that much was tolerated only impatiently by the average reader.

There were a few women writers of science fiction, but they used initials, or pseudonyms, or first names of ambiguous gender in order to hide the fact. And they carefully wrote very much the kind of stories that men did. There were even a few women editors—who carefully bought the same kind of stories that men editors did.

They weren't traitors to the cause of feminism. It was just that nothing else was possible as long as 90 percent or more of the readers were young men.

Even in the late 1950s and early 1960s, when the sexual revolution was in its beginnings, science fiction lagged behind. Then came something new!

In the middle 1960s, "Star Trek" appeared on television. It was reasonably good science fiction, several giant steps beyond the *Flash Gordon* serials. It was not the first good science fiction to appear on the screen, to be sure. I need only mention *The Shape of Things to Come* and *2001: A Space Odyssey* to make that point.

"Star Trek," however, appeared *every week*. It built a steady audience and, as it turned out, a fanatical one. When the television moguls tried to end it after its first year, the outpouring of protest was one of the phenomena of the age. For once (and perhaps only that once), the front office was frightened by the sheer weight and force of popular anger into continuing a program they thought unprofitable.

"Star Trek" continued for three seasons, has lived on for over twenty years in reruns, and has recently given birth to five major motion pictures and a new television series.

More important than that mere fact was this: For the first time, a piece of science fiction not only gained a mass following, but gained one in which women made up a large percentage. For the first time!

Why was this? My own feeling is that the makers of "Star Trek," Gene Roddenberry, in particular—did their best to build human interest into the stories, and to give the characters distinct and developing personalities from week to week. And they had Mr. Spock!

Mr. Spock was a member of the superrational, nonemotional Vulcans (though he was half-human), who maintained his calm under all conditions. For some reason (I am not a woman and I will not attempt to pontificate about the nature of these reasons), this appealed to women. What's more, Mr. Spock had pointed ears, and for some reason this, too, seemed to appeal to women.

In any case, when the time came that "Star Trek" conventions were held and thousands upon thousands of "Trekkies" swarmed to attend, half or more seemed, to my dazzled eyes, to be young women.

Those young women were ready for the next science fiction phenomenon: *Star Wars*. It was no longer an audience that was predominantly male that filled the movie houses. The young women were the first in line and were more pertinacious in repeating than the young men were.

Of all the tens of millions of viewers who sat before the television screens and flocked to the movie houses, merely tens of thousands turned to the printed media, but perhaps half of these were women.

The result was that from 1965 on we have seen the gradual feminization of the audience for printed science fiction. At the very least, 25 percent of the readers of science-fiction magazines and novels are now women. I suspect that the percentage is now nearer the 40 percent mark.

This has introduced enormous changes. For one thing, it has brought a broader scope to the writing. To satisfy reader demand, the stories have to contain women as *people.*

In addition, women writers have grown more prominent, so that Ursula K. LeGuin, Joanna Russ, Joan Vinge, Connie Willis, Octavia Butler, and others are now in the top echelons. Women are increasingly becoming editors of science-fiction magazines and books, and Judy Lynn del Rey, Shawna McCarthy, and Betsy Mitchell became powers in the business.

This feminization of science fiction is producing its rewards. Science fiction in the printed media may not be producing the millionaires that the visual media has, but there is progress. Science-fiction novels now appear on the bestseller lists. Frank Herbert, Robert Heinlein, Anne McCaffrey, Arthur Clarke, and others have all appeared there.

And the writing is better, too. It's as I always say: Liberate women, and men will be liberated as well.

# 97

# Back Through Time

Modern writers who have attempted science fiction or fantasy have, on numerous occasions, offered plots in which the hero or heroine has gone back into the past. There was, of course, Mark Twain's *A Connecticut Yankee in King Arthur's Court,* published in 1889, and in 1986 the motion picture *Peggy Sue Got Married* was released.

Why this interest in going back in time? For one thing, it enables one to have fun with anachronisms, to say nothing of the possibly hilarious confusion of the time traveler who finds him- or herself out of time and of the equal confusion of those who must deal with such a person. More seriously, it enables the writer to lend a sharper point to satire, since the protagonist (and the reader or viewer) know the future, while all the other characters in the story except for the time traveler don't.

Nevertheless, I suspect that the chief point in concocting such a story rests with the fact that human beings have a hankering to return to the past, so that tales dealing with such an event are quite likely to be popular. I'm sure that everyone of us has at one time or another felt the urge to step back into the past, even if only briefly. But before considering why that should be, let me disappoint you.

It's impossible!

Mark Twain's Connecticut Yankee traveled into the past quite involuntarily, when he received a heavy knock on the head. I'm afraid all he would have gotten as a result, in real life, would have been a concussion. Peggy Sue, in the motion picture, traveled into the past as a result of a fainting fit that may or may not have been clinical death for awhile, but all she would have gotten out of that would have been unconsciousness. Of course, it may be that in both cases the travel into the past was a dream, a flight of fancy, as it was for Dorothy when she visited the Land of Oz. Still, you can't rely on having a proper dream, so don't go banging your head against the wall or inducing a fit of some sort. Besides, a dream is cheating. What one wants is a *real* trip into the past.

In many science-fiction stories, the trip into the past is by way of some futuristic machine that can take you through time at will, as an automobile or airplane takes you through space. Such a machine was first used in H. G. Wells's *The Time Machine,* published in 1895. That, however, is totally impossible on theoretical grounds. It can't and won't be done. (If you're one of those romantics who thinks nothing is impossible, I won't argue the case, but I trust you won't decide to hold your breath until such a machine is built.)

Nevertheless, even if time-travel is not in the cards, there still remains this common yearning to go into the past. An impossible dream, perhaps, but a beautiful one.

I must disappoint you again, however. I don't think it's a beautiful dream; I think it's a nightmare. Let me explain.

One reason to want to go into the past is to experience youth again. Why not? Youth is better than old age. How wonderful to have young limbs pumping tirelessly again, in place of the creaky old body you now have. You may think longingly of the simple joys of youth, the security of being cared for by your parents, the fun of play, and so on, and so on, and so on.

But in order for this to be meaningful, you must go back into the past with your adult memories intact (as was true for both the Connecticut Yankee and Peggy Sue). If you're a youngster again, but with only the youngster's memory, you simply live your life once again, without any sense of glory in youth and health and fun. If, however, you *do* have your adult memories, you can appreciate and delight in the change—except that, knowing the future, you know what lies ahead. You know that at such and such a time you will experience a serious accident, or a terrible disease, or have someone you love die. That would make life unbearable, I promise you. It is only because we *can't* foresee the future that our lives are bearable right now.

Beside that, youth was not the pleasurable time you may think it was. Our memories are treacherous, eliminating the unpleasant and painting the pleasant in unrealistically bright colors. I took the precaution of keeping a diary from the age of eighteen, so I know this. If you kept a diary, you would, too. Take my advice and be as you are, old and decrepit though you may be. Why subject yourself to disillusion and find that your dearest memories are cobwebs of illusion?

You may decide, of course, that you don't want to go back and relive your youth. You merely want to see your parents again, your little sister, your old cronies, the old swimming hole—whatever.

Thornton Wilder did this magnificently in his drama *Our Town,* produced in 1938. The heroine, who had a chance to go back after death, simply broke her heart. She couldn't endure watching everyone live as though life were eternal, not cherishing each other while they had the chance.

The truth is even less romantic than this. Undoubtedly, you will find that your parents are not quite as you remember them, nor your family, nor your friends, nor your surroundings. Everything will be smaller, dumber, less interesting. Again for your pains, you will have disillusionment. Far from regaining youth for a second time, you will lose what you had (in memory) the first time.

But then, you may not want to go back into the past to relive your own life or to renew your old memories. You might just want to go back to a simpler time—a time before the problems of today, a time before the nuclear threat, terrorism, drugs, traffic tie-ups, pollution, and the myriads of ills society seems heir to now.

Thus, Jack Finney, in his "The Third Level," written in the aftermath of World War II, had his hero go back to the last decades of the nineteenth century, and leaves him sitting on a porch, sipping cider through a straw in the quiet twilight.

The hero is living a quiet middle-class life. But let him visit the city slums of the 1880s, and remember that before the Great Depression, the government of the United States felt no responsibility whatever for the poor. Or let him continue to sip cider until he feels the need for some entertainment, in which case he'd better enjoy a quilting party, for there were neither movies nor television to amuse him. And he'd better not get sick. No antibiotics, no modern surgical techniques. Then, too, he has to live with the sure knowledge that the twentieth century is coming with its world wars, fascism, communism, and everything else. Restful? I think not.

But hold on, suppose you're not going into the past merely to relive a simpler or more youthful time in a passive manner. Suppose you're going back for some active point. You are going to *change* your life. You're going to find the fork where you took the wrong turn (haven't we all gone wrong at some time or other?) and change it. (In the drama *Morning's at Seven,* there was one character who constantly leaned against a tree, saying, "I've got to find the fork.") Peggy Sue, for instance, finding herself back in the past, is determined not to marry the very unpleasant young man she had married, because the marriage had proved to be an unhappy one.

However, taking the other road when you come to the fork is not merely going to change one consequence and leave everything else unaltered. The other road is going to lead to innumerable unforeseen consequences, while the first road, left untrodden, is going to remove equally innumerable consequences you might not want removed. For instance, to take a very simple example, if you decide to wipe an unsatisfactory husband out of your life, it may be that he provided you with a child you adore. That child gets wiped out with your husband, of course, and you have to weigh the benefits against the harm. In fact, for all you know, taking the other road may lead to an agonizing death for you the next day. How can you dare touch anything?

Ray Bradbury made that point marvelously in his "Sound of Thunder." Hunting safaries go back into the Mesozoic era to track and possibly kill dinosaurs. This is done in strictly restricted areas and under restricted conditions calculated to prevent any change in the future. One person carelessly steps outside the bounds and unintentionally kills a butterfly. The present world is enormously changed as a result.

Or perhaps the time traveler doesn't want to change his or her life but merely use his foreknowledge to become wealthy. Thus Peggy Sue discovers that way back in the dark ages of 1960 there was no pantyhose. There is a scene, then, where Peggy Sue is shown sewing pantyhose, perhaps with the intention of making herself rich, rich, rich. This is not followed up, but it is not likely that the mere existence of one pair of pantyhose is going to lead to wealth. It may seem, after the fact, that there is an overwhelming demand for pantyhose, but that demand had to be *created* by publicity and advertising campaigns, which would require a great deal more capital than Peggy Sue would be likely to have.

The same probably goes for any other get-rich-quick scheme the time traveler has. One may know the next Derby winner or what the stock market will do, but those first few winnings will undoubtedly change reality in such ways that one's foreknowledge is wiped out.

The time traveler might, however, be a true idealist. He (or she) might not care what happens to himself as a person. He may be intent on changing the world for the better, and damn the particular consequences.

Thus, it was the Connecticut Yankee's dream to introduce nineteenth-century American initiative and technology into King Arthur's court, thus wiping out slavery and the backward mummery of Merlin (the villain of the story). Again, in L. Sprague de Camp's "Lest Darkness Fall," the hero, Martin Padway, goes back to Ostrogothic Italy just before Belisarius's campaigns wipe out what is left of Roman civilization in the West and introduce the "dark age." Padway strives manfully to introduce twentieth-century technology and prevent this dark age.

Would such grandiose schemes succeed? Personally, I think not. It is very unlikely that one can graft the technology of one century onto the social system of another. In other words, "you can't have steam engines until it is steam-engine time."

Besides, what may seem a desirable change to one person may not necessarily seems so to another. There have been a number of stories written of people who go back into the past and seize an opportunity (or plan one from the beginning) to prevent the assassination of Abraham Lincoln. Naturally, such a scheme of prevention always fails.

On the other hand, some other person finding himself a few years farther in the past may be an ardent Confederate patriot and may exert himself to bring about the assassination of Abraham Lincoln before he took office.

Would the former change have really prevented the ills of the Reconstruction period? Would the latter change have really led to the Union's loss of the Civil War and the establishment of an independent Confederacy? Do we really know? And what other consequences would follow in either case? *Do* we really know?

In short, going back into the past is not going to help anyone. Whatever a time traveler's purpose is likely to be, he or she will surely be disappointed.

We may continue to dream on (there is no charge for dreams), but let us be glad, then, that time travel is impossible.

# Part VII
# Personal

Finally, I include four essays that are rather intensely personal. The first two have the joint byline of Isaac and Janet Asimov, Janet being my dear wife.

Janet is a psychiatrist and psychoanalyst by profession, but since her adolescence she has dreamed of being a writer and, in recent years, especially after she retired from active practice in 1986, she has been working at her science fiction. (It was our mutual interest in science fiction that brought us together.)

So far, she has published fourteen books and has two more in press. Five are under her own name, either Janet Asimov or (her maiden name) J. O. Jeppson. The two most recent of these are *Mind Transfer* and *The Package in Hyperspace,* both published by Walker in 1988.

The remaining nine books, and the two in press, appear under our joint names. These include a series of books about a small and lovable robot named Norby, an anthology of humorous science-fiction stories, and a nonfiction book entitled *How to Enjoy Writing* (Walker, 1987). Wherever our names both appear, Janet and I discussed the book and then she did the writing and I did some unimportant patchwork.

And that's the way it was with the first two essays in this final section. Almost all the writing is Janet's.

The remaining essays are autobiographical, and the last, in particular, was inspired by the fact that on May 3, 1987, I was declared a "Grand Master" by the Science Fiction Writers of America—a real climax to my career and much appreciated. I wrote the essay "Seven Steps to Grand Master" in a humorous vein, but don't let that fool you. The award made me *very* happy.

# 98

# Our Shangri-La

## by Janet and Isaac Asimov

Most people love to take trips, but in our family it is Isaac who makes most of the rules, and his idea of adventurous traveling is to take a fifteen-minute trip by taxi to an editor in midtown Manhattan. It is perhaps astonishing, then, that, without a murmur, he is willing to accompany Janet to one exotic place we visit several times a year—Mohonk Mountain House!

We think of Mohonk as a nonfictional Shangri-La, a far-off place reached only by a journey of both body and mind, a place of beauty and kindness, where time does not have its usual deleterious effects. Mohonk, like Shangri-La, is hidden, not in a secluded Himalayan valley, but in the folds of a stony mountain ridge formed some 450 million years ago.

We prepare for the two-hour car trip from Manhattan as if we were venturing through primitive wilderness. We have maps and thermos bottles, first-aid kits, umbrellas, boots—and reading matter, in case we are delayed by road construction or some other untoward event in the wilds of suburbia.

We lock up the apartment, then unlock it and go back inside to make sure that we have turned everything off (especially our typewriters and word processors, since we both write for a living). Then we lock up again and hesitate while we wonder if we have really taken care of everything. Then we decide to be carefree daredevils and to assume that all is well and boldy advance to the elevator.

Once the elevator door opens, Isaac's heart usually fails him and while Janet holds the door against its petulant desire to close, he rushes back to the apartment to make sure that he really locked the door instead of just standing there working out some detail of the plot of his new novel. Janet assures him that she distinctly heard the locks click, but he merely sneers at hearsay evidence.

And then we dare the thirty-three flight elevator trip and trudge three further flights downward into the sub-basement where our car is parked. It invariably starts (to our great relief, but then it's a Volvo) and we drive to the West Side highway, where other specimens of *Homo sapiens,* who drive over it every day and therefore lack our thrill of adventure, whiz past the boat basin, the cherry blossoms (in season), and the brand-new garbage disposal plant as if they were not there. We observe it all on their behalf.

Across the George Washington Bridge (over that large sea-inlet known as the lower Hudson river) we drive right through a cut in tilted lava, the great columnar basalt of the Palisades, to the Palisades Parkway, bordered by flowering shrubs, trees, and glimpses of the Hudson. This takes us to the Thruway, which heads toward the Ramapo hills, but veers off into the Palisades Interstate Park, climaxed by Bear Mountain.

We persevere. The first glimpse of the distant Catskills makes us look to the left, over apple orchards, to a high rocky ridge—the Shawangunk Mountains, weathered down because they are one hundred million years older than the Catskills (which aren't a folded mountain ridge, but only an eroded plateau.) Perched on the top of the Shawangunks is Skytop tower, and hidden just below it is Lake Mohonk.

We turn off at New Paltz—with its two-hundred-year-old Huguenot stone houses, as well as a modern college—promise ourselves that we will stop at the bookstore later, and drive on, across the Wallkill river, past cornfields. If we take the correct road, we head up the ridge to the Mohonk gatehouse, where we're checked and waved on.

Then the spell of Mohonk begins, as the road narrows and climbs, overarched by trees that seem part of an endless forest. As our ears pop (who needs to climb Mt. Everest or go up in a plane?), we put the car into second gear and go upward, seeing through the trees the amazing view across the Rondout Valley to the Catskills. In mountain-laurel time, white flowers are massed among the trees, and in autumn the dark green of the hemlocks sets off the display of color elsewhere.

The road passes through a gap in a stone wall, curves around (icicles dripping from the rock cliffs in winter), and suddenly there is the Mountain House, an almost Charles Addams creation bursting with bays and turrets and balconies. We say, with immense relief, "It's really there!" and slow the car so that we can admire the enormous copper beech beside the road, and then the spectacular formal gardens. We stop the car under the stone arch that covers the front driveway and enter our own Shangri-La.

We write about other worlds in our separate forays into science fiction; Mohonk Mountain really is one. Except for improvements in such things as plumbing, it manages to look much the way it did when Janet's parents visited it forty and fifty years ago: full of oak furniture, incredible potted plants, ancient pictures, nooks and crannies, and a window with hummingbirds sipping from a feeder just outside.

Our room will have a view of the Catskills or—on the other side—a view of Lake Mohonk with Skytop tower on the white quartz-filled cliffs of Shawangunk Conglomerate (from the Silurian), resting on dark Martinsburg Shale (Ordovician). There are eons of Earth history before our eyes, and we begin to feel very young in comparison. Even the drinking water tastes good. Maybe they've tapped into the Fountain of Youth.

At one time or another, we've enjoyed most of the various holidays at Mohonk, and many of the special programs, including the December science-fiction weekend, which Isaac has been persuaded to preside over.

Perhaps what we like best is Music Week in June. The chamber music concerts are in the Parlor (one always capitalizes it and is grateful that there is no smoking allowed there), and the acoustics are perfect. There are carved wooden walls and ceilings, transoms, ceiling fans, windows with intricate panes, an immense fireplace, huge oriental vases, and even an old Metropolitan Opera chair.

The Smiley brothers and their wives, who started Mohonk in 1870, look down from portraits on the walls, and their descendants are with us in the hall, because the Smiley family still owns and runs Mohonk. We go to the morning chapel and sing along with the organ as it is played by Mrs. Smiley, who is, theoretically, in her nineties, but whose agelessness makes us feel hopeful about the future. Now that we ourselves are past our first youth, we need all the hope we can get.

Of all the manifold attractions of Mohonk, Isaac likes the meals best. He's funny that way. A special attraction: there's no smoking in the Dining Room, either. He likes breakfast, sometimes with a view of clouds filling the Rondout Valley like a white sea lapping against the Catskills. Or the buffet table at lunch. Or dinner with the lights reflecting in such a way that they seem to hang outside, too.

Then there's teatime in the Lake Lounge, which we never miss, and where Isaac can be counted on to steal another cookie whenever Janet goes out to buy fish food with which to feed the trout that wait off the porch that overhangs the lake.

We don't play tennis or golf or hang on skinny ropes from cliffs, white Conglomerate or not. Isaac prefers to write in our room while waiting for the next meal (he's far more compulsive about his writing than Janet is). As for Janet, what she likes to do is to walk around Lake Mohonk or up any of the myriad trails that surround the Mountain House (we have yet to explore all of them). And since Isaac loves Janet in his writing-preoccupied way, he generally accompanies her, although sometimes with an unlovely air of being martyred. Fortunately, Mohonk not only has the best walking trails in the Universe but plenty of seats to sit down on in case you want to admire the view, to rest, or (in Isaac's case) to ponder upon a plot.

Isaac, in particular, has learned a lot at Mohonk. He's discovered that he's almost not afraid of horses when they're pulling a carriage. He's not afraid of being on a lake in a pedalboat (provided Janet stays near him), even though he can't swim.

One evening we were favored with a sight that almost made us forget human greed and destructiveness. In the course of a stroll, we met nine deer, who stared at us placidly from thirty feet away and went on feeding. And once we saw a single doe, not ten feet away on one of the mountain trails, who didn't even bother to favor us with a look. They did not fear our cruel and vicious species, and we were both moved and flattered.

One year we lay on a grassy lawn for half the night, face up, watching the Perseid meteors flash by while Mohonk turned off its lights to improve the viewing. We've attended mystery weekends and toured the strange and eerie cavernous and interminable cellars of the house. (Fortunately, Isaac is as brave as a lion, and held on to Janet's hand, tightly, only in case *she* was scared, of course.) We've watched an outdoor show of a filmed *La Traviata* in color (with subtitles) and indoor showings of old black-and-white classics.

Some years we've stood at Cope's Lookout, seen gypsy moth damage on distant Shawangunk ridges, and thought about Dan Smiley's philosophy of doing without pesticides (because the ecology of Earth will manage itself, if given a chance). (Isaac, the chemist, is not entirely confident of this.) At Mohonk, however, it seems to work. At Mohonk, the interdependence of living things is obvious and reverence for life is easy.

And sometimes we just spend a couple of days of detachment from the cockamamie world of so-called reality, and drift in a dream world where the Fall has not yet taken place and where Eden still exists.

Janet says that Mohonk instills in her a Zen acceptance of a reality not lacking in truth, beauty, and love, an acceptance mixed with the Quaker joy inherited from the Smiley family. So she always leaves with an apple tucked in her purse as a last reminder, while Isaac clutches the handwritten manuscript of his latest story—after we have carefully made a reservation for the next time.

# 99

# The Tyrannosaurus Prescription

## by Janet and Isaac Asimov

"Escape" is a word New Yorkers must take seriously. Most of us talk about it to each other, and Outsiders (that is, people who, for some odd reason, *don't* live in New York) are always asking us, "Why don't you escape from the city?"

"We like it here," I mumble, bracing myself for the forthcoming lecture. The Outsiders talk about all the things they had to run about doing while in the city, things that exhausted them and, more importantly, their wallets (things we quiet-living inhabitants hardly ever do); and they clearly hold it all against us. They go on to list the deficiencies of urban life, especially urban

Manhattanite life, such as crime, crowds, noise, air pollution, litter, high taxes, gridlock, water mains that must have been put in by Peter Stuyvesant on a bad weekend, and parking garages that charge a pint of blood per hour.

"Move out," they say, peremptorily, "or at least buy a country weekend retreat."

Janet, who in her time has lived in countrified areas, says politely, "But then we'd be escaping to neighbors' power motors, to surprise visits with no preliminary phone calls, to recalicitrant septic tanks—and we'd have a long haul to Lincoln Center."

To which I add, "Plus having to live with birds that start soprano conversations early in the morning, when I'm trying to write."

That just gets me a scowl from Janet, who is of the opinion that bird conversation is pleasant, so that she opens our east windows at dawn to hear the squawking from Central Park. (Well, she says "warbling.")

"And you should travel," say our friends. "That's the best escape."

We've noticed that their homes in the country don't stop *them* from traveling. They seem to want to escape the joys of nonurban existence in order to travel to far-away lands, where they can experience crime, noise, crowds, air pollution (and water pollution, too, which we *don't* have in Manhattan), high prices, dysentery in its various forms, and (I admit it) less litter.

So I say, firmly, "There's plenty of escape right here in New York. I don't even count work. In my own case, as long as I have paper and some machine that will help me put words upon it, I can escape into my own head whenever I please, but I won't count that. Consider instead. . . ."

By this time, I have usually lost my audience, but who cares. *I* know about escape and it comes under the most apparently mundane conditions. When we visit the new quarters of my publisher, Janet (who has decided to be an amateur geologist and paleontologist) points out to me the fossil shells in the polished limestone walls at the main entrance. For a few minutes we are not in Manhattan, but floating in warm primordial seas.

Living in New York provides endless possibilities for such moments that involve not only the inanimate environment, but people as well. There is nowhere in the world where you can watch the passing parade of all the peoples of the world as you can in Manhattan.

And nature, too, believe it or not. Trust Janet for that. She can sniff out nature in the unlikeliest places, but just so that she won't have to, she managed to get us an apartment overhanging Central Park.

The result is that, however ostentatiously busy I am keeping myself, I am sure to hear those fateful words, "Come for a walk."

It doesn't do me any good to pretend I've heard nothing, for Janet will then say in severe tones, "You *must* exercise something other than the tendons of your fingers."

And she's right, I suppose. I must say that I prefer sitting to walking, but Janet, for some reason, does not, and on a clear vote of 1 to 1, she usually has the majority.

But then again, Manhattan is a walker's city. The traffic being what it is, the fastest and best way of getting from point A to point B is to lift and advance your two legs in rhythmic alteration.

Janet's idea of an ideal outdoor walking escape is into the wilderness (which she takes to with all the vigor of her Viking ancestry). I prefer sidewalks myself, but I will confess that there is an extraordinary amount of nature in New York—zoos, botanical gardens, roof gardens, even a remnant of primeval hemlock forest, Janet tells me—but we don't have to look for any of it ordinarily. We just cross the street and throw ourselves into Central Park.

There's nothing tame about Central Park, and I'm not referring to all that loose talk about muggers and crime. Central Park is full of ponds and lakes, hill and glacier-scarred rocks, gardens, and even the wild Ramble, where migrating birds stop to rest after having turned up their beaks, haughtily, at the less inviting country areas outside the city. You can see anything from a woodchuck and her babies cropping the lawn outside the Children's Zoo to people whirling about in one of the nation's classic carousels. Lunch at the boathouse makes me feel I am lost in a strange, sylvan world—and I really am lost, for I have no sense of direction and must rely on Janet to guide me home.

Our enjoyment of Central Park is deepened by knowing that Frederick Olmstead designed and built the park from a treeless, garbage-strewn wasteland filled with shanties, goats, and fierce squatters. Olmstead had to have an armed guard when he surveyed the territory.

Nature exists in the city streets, too. We have seen sparrows dive-bombing a blue jay that was trying to get at the eggs in a nest inside a lamp post above a busy sidewalk. We were the only people looking up to watch the small-scale life-and-death situation.

Janet also walks in museums, and sometimes I go with her. At that soothing oasis called the Frick, I sit by falling water in the covered courtyard, listening to the next few pages I must write reeling themselves off in my head, while she communes with the *Polish Rider,* which may or may not be by Rembrandt, but which resembles (she confides to me) Yul Brynner.

We also hang out in the Metropolitan Museum of Art, further up Fifth Avenue, where there are more Rembrandts. Janet has a patent crush on *The Noble Slav,* so I do my best to emphasize my own Slavic cheekbones while trying to look noble.

Of course, New York consists of thousands of neighborhoods, each its own distinctive milieu. Our own couldn't be improved upon. Besides Central Park, there is Lincoln Center, the New York Society for Ethical Culture, and many good restaurants (most of them deliciously ethnic—*I* am the expert on that). Then there is my bank, where I put in a little more than I take out (Micawber's prescription for happiness, and it does make the bank happy). There is the butcher, who tolerates Janet's weekly purchase of a small piece of liver (for medicinal purposes). And there is the post office, where I drop my manuscripts into the slot. Let them travel even if I don't.

Then there is a mystery bookstore not too far north, and a general bookstore even less far south. Janet and I grew up going to libraries and longing for our own books, and now that we can afford them in riotous plenty, bookstores mean more to us than catnip to a cat. It's nice having a wife who prefers bookstores to Bloomingdale's.

But what do we do when life's viscissitudes bear down at us most heavily, when worries accumulate, when the word processor is misbehaving, when Janet is troubled over the fate of the world's rhinoceroses, when I have reached a cliff-edge in my novel and don't know what's coming next, when Janet becomes conscious of a new and disturbing noise from outside, when we both grow concerned at the hole in the ozone layer? What then?

"We've got to do something," muses Janet.

And I know exactly what to do. Experience has taught me. It's time for the Tyrannosaurus Prescription.

"Uptown!" I say. "Get your walking clothes on."

So it's eleven blocks north to the American Museum of Natural History. Janet flashes her life-membership card, and we rush over to see "October at Stissing Mountain" so that we can relive a trip we once took along Route 17 at the peak of the autumn foliage season.

We revisit England's New Forest, with bluebells under the three-hundred-year-old beech trees. Joining the cougars in their cave, we look out on Grand Canyon, which (Janet tells me) looks just like the real thing except that there's no haze-pollution obscuring the view.

To rest, we sit beside the African elephants, a herd that is helping a wounded comrade, and remember, defensively, that human beings are sometimes altruistic, too. From our seat, we can see the gorilla exhibit, and Janet tells me how Carl Akeley planned the hall and invented the method of sculpting animal shapes instead of merely mounting skins.

And for the last and best, I march her to the elevators. We ascend to the fourth floor and there he is, forty-seven feet long, 18.5 feet of him, with a four-foot skull and big sharp teeth. *Tyrannosaurus rex,* king of the master-lizards, the largest of the land carnivores, and not intelligent enough (or stupid enough) to devise nuclear bombs and then to continue stockpiling them when enough already exist to kill every person on Earth twenty times over.

We like *Tyrannosaurus rex.* He and his kind left Earth without ruining it in the process, and—who knows—perhaps, in the end, we will, too. Humanity is fiercer than he is, for all his teeth, but perhaps we will yet be wise enough to turn our fierceness on our true enemies—ignorance and misery.

And we laugh at the thought, and feel better. The Tyrannosaurus Prescription has worked and our sense of humor is back—and that is the best escape of all.

# 100
# Ellis Island and I

In 1922, I was two years old, and I lived with my mother and father in Petrovichi, on the eastern border of Byelorussia in the Soviet Union. My family had survived World War I and the Russian Revolution and the Civil War and the Intervention. Things were now relatively peaceful, and we were reasonably well off.

However, my mother had a half-brother, Joseph, who had emigrated to America before the war and who now wrote to Petrovichi to inquire after the welfare of his sister. When my mother answered with reassurances, he wrote again and offered to sponsor our emigration to the United States.

My parents considered the matter and decided to take him up on the offer. They were not fleeing oppression or anything of the sort. It simply occurred to them that it might be a move for the better. Nor did they have to leave surreptitiously. They applied for a passport, received it with no more than the usual routine bureaucratic delays, and off they went—taking me with them as a matter of course.

On January 11, 1923, shortly after my third birthday, we left the Soviet Union, and, as it happened, we never returned. The sea voyage, I was told later (naturally, I remember none of it) was unpleasant indeed. While traveling from Riga, Latvia to Liverpool, England, my father was constantly seasick. During the next stage, from Liverpool to New York, it was my mother's turn. Neither of them had ever been at sea before—and they never were again.

On February 3, 1923, the ship (the *Baltic*) sailed into New York Bay, passed the Statue of Liberty, which greeted us silently as part of the "wretched refuse" that had streamed in by the hundreds of thousands during the third of a century it had been standing there, and docked at Ellis Island.

My father walked down the gangplank with me in his arms, utterly confused by everything that was going on, by all the orders shouted at him in incomprehensible gibberish. It was only later that he noticed that my mother was gone. The men and women were separated and he didn't see her for four days. I increased the joy of the occasion by coming down with measles (which may have delayed the reunion).

Once he had gone through all the routine, the medical examinations, the questions as to how he planned to support himself ("By any honest work I can find to do," he answered, surprised that anyone should think it necessary to ask), he found my mother, with her brother, Joseph. And, on February 7, we were left to our own devices in the new country, which my parents were never to leave. Fifteen months later, new immigration laws established strict quotas against East Europeans, and we would undoubtedly not have been able to get in.

I remember nothing of Ellis Island. I revisited it many years later, when it was an abandoned wreck, but no spark of memory was stirred.

I remember well, however, the first place we lived at after having come to the new land. It was at 425 Van Siclen Avenue, between Sutter Avenue and Blake Avenue in the East New York section of Brooklyn. It was so primitive a place, it lacked electricity, and once darkness fell we had to light the gas jets. My parents were used to that, however, so that we didn't suffer unduly.

For my parents to go from a small "shtetl" in western Russia to the enormous city of New York was very much like going from Earth to Mars. The simple scale of the city struck my father dumb with astonishment. He tells me he stood for a long time watching successive trains pass by on the elevated, and wondering where they obtained all the people to fill all those cars.

Culturally, my parents had changed from knowledgeable and educated people into humble illiterates in the space of a month. In Russia my father read, wrote, and spoke Yiddish, Hebrew, and Russian, all fluently. He had a thorough Talmudic education and was intimately acquainted with Russian literature. He and my mother engaged in amateur theatricals with great success, and my father ran a library in Petrovichi and read Russian books to those who could not.

In the United States, to be sure, we lived in a Jewish neighborhood, where Yiddish was understood by everyone, and Russian by many, but the dominant language was a complete mystery. Even the Latin alphabet was confusing to someone who knew only the Hebrew and Cyrillic alphabets, so that at Ellis Island my father managed to misspell our name, making it "Asimov" instead of "Azimov" through a misunderstanding as to the nature of the sound of the letter *s*. (Never mind, I much prefer Asimov.)

The signs in the store windows, the street signs, all directions, were strange and incomprehensible, so that my father hastened to those signs that were in Hebrew lettering, if only to convince himself that he could still read. Years later, he told me that his first attempt landed him at a glazier's shop window and, the sign said, in Hebrew letters, "Vindez gefixt." Thunderstruck, it suddenly occurred to him he couldn't read Yiddish either!

No longer the cultural leader of a town, my father, who had worked for *his* father (a well-to-do merchant) as an accountant, now could not find any work to do but the most menial. He attempted several jobs as a door-to-door salesman (without English). He worked at knitting machines in a sweater factory. Finally, in 1926, he had managed to save enough money for a down payment on a small candy store. He remained in one candy store or another for the next quarter-century and it was a blessing. It meant that the whole family had to work sixteen hours a day, seven days a week (well, we did take off Yom Kippur when we could), but it kept us going through the Great Depression without missing a meal.

This economic, social, and cultural demotion certainly couldn't be looked upon as the "move for the better" that had motivated my parents' decision to go to the United States, but my father had never been so naive as to believe that in America the city streets were paved with gold.

Indeed, I remember, as a young child, listening to my father tell me the story of a freshly arrived immigrant who noted a twenty-dollar gold coin on the sidewalk and who refused to take the trouble to pick up such a small piece of gold.

"You can imagine his feelings," said my father, solemnly, "later on, when he found there was no more gold in the streets."

I laughed, but my father didn't even smile. He did not think such anecdotes were meant for laughter. They were cautionary tales designed to promote virtuous and rational ways of life. The tale of the immigrant was meant to inspire prudent behavior in me and to warn me off wild and unrealistic hopes.

Indeed, it was always clear to me that my parents made their move in the hope that their children, not necessarily themselves, would be better off. That, at least, worked out. My father was dreadfully disappointed when I failed to gain entrance to medical school, but he lived long enough to see me become a professor and a very successful writer.

Nor did my father's self-image suffer. He was a man of strong convictions, who had spent his formative years in a full appreciation of his own intelligence and worth, and he refused to change that simply because of altered circumstances. I learned from him. When I failed to get into medical school, I simply changed career-direction without missing a beat, never once altering my perception of myself as a sure bet for success.

That came gradually, of course. As a child, I had to go through a period of adaptation. I remember very clearly (I must have been four at the time) being caught by my needs, and urinating at the street curb (as I may have been taught to do in Petrovichi). I was puzzled by the reaction of the woman near me, but I gathered she was displeased, and I never did it again.

I was also bothered by being called a "greenhorn," as my parents must also have been. Looking back upon it, I feel a littler bitter over the fact that the expression was used most often by people just like ourselves who had merely gotten off the boat a few years earlier.

Somehow I feel that my father and I were both bothered, most of all, by his lack of English. My father labored to learn it, struggling with English-language newspapers, but he was twenty-six when he arrived in the United States, and all his life he spoke English with a strong accent.

As for me, my father was determined that English was going to be my first language. He absolutely refused to expose me to Russian, though he and my mother sometimes spoke it when there was something they wished to discuss that was none of my business. I never learned Russian, in consequence. I regret that, for a bit of Russian would be so useful now, but, on the other hand, it did mean that English is my language without competition, and I cannot bear the thought that it might not have been. Think of my ultimate profession. (Of course, I could not help but speak Yiddish, which I was speaking when we arrived, and I still speak and understand it today—more or less.)

My father's failure to speak English sharpened my curiosity about the language, and I tried to read what he could not read—obviously without his help. I plagued the bigger kids on Van Siclen Avenue to tell me the names of the letters and how to write each one and how to pronounce it. They helped me with poor grace, but did so—probably to get rid of me.

Gradually, I learned to sound out the letters and to discover that they made up words. When I could see that the sign that said V-a-n S-i-c-l-e-n obviously stood for *Van Siclen,* the street I lived on, I was overjoyed. It was like finding the key to a box full of candy.

To this day I remember the sudden surge of triumph when I realized there must be such things as silent letters. I was on an elevated train with my mother and I amused myself by trying to sound out letters. The sign on the window just opposite my seat was C-O-N-E-Y I-S-L-A-N-D. I carefully pronounced it "sohnee issland," and that made no sense. Nor could my mother help me, for she couldn't read it either. But then, in a moment of insight, I remembered that the kids frequently spoke of someplace called "konee iland," and all at once I knew what the sign said. At one and the same time, I realized not only how Coney Island was spelled and that the second word had a silent *s,* but I also knew how people who took trains knew where they were going.

Eventually, before I had started school, I could read after a clumsy and stumbling fashion and demonstrated that to my father. My father was astonished and asked me how I had learned, but I didn't know how to explain it to him. "I just did," I said.

So he got me a small dictionary. He didn't tell me that the words were arranged in alphabetical order, perhaps because he wanted me to puzzle it out for myself. Fortunately, I did.

My father, always ready to boast about me, told my Uncle Joe that I could find any word in the dictionary. My uncle stared at my five-year-old self, skinny and small for my age, and said "Impossible!" I thought that was the word he wanted and promptly found it in the dictionary and showed it to him. I don't think he was pleased.

Once I started school, I became aware of the fact that we didn't have books at home. (We had no money to buy them with.) I expressed my unhappiness over that, and my father managed to get a library card for me even though I was only six years old at the time.

From then on, for the next fifteen years or so, I trekked regularly to the library nearest to home base and read indiscriminantly, since my parents did not know English literature well enough to guide me. In one happy period, when we lived right on the border between Brooklyn and Queens, I managed to wangle a card in both borough libraries. Then, when I was eleven and entered the ninth grade, I insisted on an adult card as a high school student, even though the librarians, judging me by my appearance, clearly thought I was no adult and did their best to refuse me.

Looking back, it seems to me that I might argue that I had to overcome mountainous difficulties to get myself an education. (School was not where I got my education. Every semester I routinely read all my new school books in the first week, and there was nothing the teacher could tell me after that.)

Those difficulties weren't really in the way, however. It was those difficulties that I *needed.* It was my consciousness that we were "greenhorns," and that my father couldn't read English, that drove me madly to correct that shortcoming in myself. It was my father's consciousness of the same thing that drove him to encourage me in this.

Only in one point did we clash in this matter of reading, and that was over the newsstand in the candy store. I wanted to read the magazines and my father was unalterably opposed. He felt that I would be reading trash and contaminating what he obviously was beginning to think was going to be a first-class mind.

For a while all my arguments fell on deaf ears, and then I discovered science-fiction magazines, which I took surreptitious peeks at while my father was taking his afternoon nap. In particular, I found one called *Science Wonder Stories,* and I pointed out to my father that since the stories were about science, they were bound to be educational.

It was a good time to attack, for my mother was pregnant with what turned out to be my younger brother, and my father was feeling as though he had a lot more on his mind that questions over whether I could read a magazine or not. He gave in.

That started me, at the age of nine, on my career as a science-fiction reader. By the time I was eleven, I felt that I just could not get enough science fiction from the magazines (there were only three, and they came out only once a month), and it struck me that I might write my own.

I didn't quite write science fiction at first, but I managed to get to it when I was fifteen, and by the time I was eighteen I sold a story to one of the magazines and was off and running.

I cannot say how things would have been for me had I not come into the United States as an immigrant. I can't go back and live life over under changed circumstances. Still, as I think about it, it seems to me I needed something to rise above.

To be brief, I'm glad I came here—and I'm glad I had to come here. Life might have been too easy for me if my ancestors had beat me to the punch and had come here on the Mayflower.

# 101
# Seven Steps to Grand Master

### Step 1: I Take an Ocean Trip

I was born in Russia. The land had just gone through World War I, a revolution, a civil war, and foreign intervention. To inflict myself on the nation at such a time was rather merciless of me, but I plead not guilty. My parents made me do it.

In late 1922 my parents decided it might be a good idea to emigrate to the United States. They were getting along, they were not in dire poverty; they had not suffered unduly as a result of the troubles the land had been suffering; but they suspected that in the long run they might be better off in the United States.

One of the problems they had to face, I imagine, was whether to take me along. I was not quite three years old, and what with contracting double pneumonia (after falling into a nearby pond and, after some hesitation, being pulled out by my mother) and inflicting other joys of the sort on my parents, I imagine they felt they might be better off in the United States by themselves.

However, largely because (I suspect) they could find no one foolish enough to take me off their hands, they sighed and put me in a knapsack so at least they wouldn't have to waste a ticket on me. I crossed the ocean with them, and we arrived in Brooklyn in February of 1923, a little past my third birthday.

This was my first step to Grand Masterability. Had I remained in the USSR, I dare say I would have received an adequate education and would have taken to writing and would even have begun to write science fiction (making use of the Cyrillic alphabet, to be sure), but I don't think things would have gone as well for me there as here. In 1941, the Nazis invaded the Soviet Union. I was twenty-one at the time, and I suspect I would have been in the fighting and might well have been killed or, worse, been taken prisoner. Or, if I had survived, I might conceivably have gotten in trouble with the regime because of my tendency to speak out of turn. (I have frequently gotten in trouble in the United States for that reason, come to think of it.)

And, finally, I don't know if they have a Grand Master award in the Soviet Union. So, all in all, that ocean trip was essential.

### 2: I Insist on My Identity

Once we were in the United States, my parents realized that they had gained a new status: they were "greenhorns." Everyone was eager to advise us and to guide our faltering steps down the pathway to American citizenship, especially those old settlers who had gotten off the ship five years earlier.

One neighbor woman said to my mother, in my hearing (I was four or five by then and so small for my age that no one noticed me—so that I got stepped on a lot), "Why do you call him Isaac, Mrs. Asimov? With a name like that he will always have a stigma on him." (Translation: "Everyone will know he is Jewish.")

My mother said, "So what should I call him, Mrs. Bindler?"

And Mrs. Bindler (or whatever her name might have been) said, "Call him Oiving." (Translation: Irving. This is a grand old aristocratic English family name.)

My mother was very impressed and would undoubtedly have accepted the suggestion, but, as I said, I was listening with each ear. I was not yet old enough to understand the semantic fact that the name of a thing is not the thing itself. I didn't understand that I was merely *called* Isaac and that I could be me whatever I was called. (Or as I once put it—rather neatly, I think—"That which we call a rose by any other name would smell as sweet.")

What I thought was that I *was* Isaac and that if I were called anything else, I wouldn't be me. Whereupon I raised what we called in those days "a holler," absolutely refusing, under any conditions, to allow myself to be called Oiving. I was Isaac and I intended to stay Isaac. And I did. My mother simply wilted under the force of my indignation.

Without knowing it, that was my second step toward Grand Masterdom. Had I accepted Oiving, it would have proved every bit as stigmatical as Isaac, for so many Jewish mothers had sought escape for their young hopefuls in that direction that Oiving became as Jewish as Isaac and without the biblical cachet of the latter name. (Besides, Newton's name was

Isaac, too, and that's an even better pedigree than the Bible as far as I'm concerned.)

Having escaped Isaac, I would have ended by despising Oiving and would have changed my name to Ian. Then, realizing that Ian went but risibly with Asimov, I would have changed my last name to Ashford, and it would have been as Ian Ashford that I would have written my science fiction.

Now, I am a strong believer in the value of name-recognition. No one would have noticed or remembered a name like Ian Ashford. However, the name *Isaac Asimov* attracts notice at once. People laugh and have long discussions over how it might be pronounced. When a second story appears with the same name, they see it again, and before long they can hardly wait for another story by me. Even if the story is no good, the name makes a terrific conversation piece. I would have sunk without a trace if I had not had the good sense to keep my name.

**3: I Live on the Subway Line**

I was eighteen years old and I finally had a story I wanted to submit to John W. Campbell, Jr., the new editor of *Astounding Science Fiction.* The trouble was, I did not know how to do that. The logical way was to mail it to him, but the story, plus envelope, weighed just over three ounces, which meant four three-cent stamps, or twelve cents altogether.

If I went by subway, it would be five cents each way, or ten cents altogether. Of course, the subway would mean half an hour of my time each way, but in those days my time was worth nothing. Weighing the relative values of twelve cents and ten cents, I came to the conclusion that two cents was valuable stuff. I therefore took the subway.

Approaching the receptionist in an agony of fright, I asked for Mr. Campbell, expecting to be thrown out, with the manuscript following me, divided into four pieces per page. Campbell was willing to see me, and we talked for an hour. He gave me a quick reading and sent me a quick rejection with a very kindly and helpful letter. After that, I visited him once a month and I was on my way.

How did that happen? What was the deciding factor?

Easy! I lived half a block from a subway station. Had I lived in Fargo, Nebraska, the railroad fare would have been more than twelve cents. For goodness's sake, if I had lived in Staten Island the Ferry would have added ten cents to the round fare, and comparing twelve cents and twenty cents, I would have put the envelope into the letterbox. I would then never have met John Campbell and I would not have received the kind of encouragement and charisma that poured forth from that great editor.

Hurrah for the subway station! I might never have made Grand Masterhood without it.

## 4: I Walk in at the Right Moment

I tried to come to Campbell with a new idea every time I saw him. Every once in a while, though, Campbell had an idea of his own. In the competition between a writer's idea and one of Campbell's ideas, it was Campbell who always won—at least when it was I who was involved.

One day, Campbell had a terrific idea, and he was aching to force it on some writer. He never told me the details, but the picture I have in my mind is of Campbell sitting there like a vulture waiting for an innocent writer to enter his lair (assuming vultures have lairs)—*any* innocent writer.

It must have come as a nasty shock to him when I, aged twenty-one, and just as innocent as they come, walked in and said, "Hello, Mr. Campbell." It's undoubtedly a tribute to the manner in which that idea had him in its grip that after a momentary shudder, he dismissed the idea I was trying to describe and said, "Never mind that, Asimov. Let me read you this quote from one of Emerson's essays."

He read it—something about how if human beings could see the stars only once in a thousand years, they would get a big kick out of it.

"They wouldn't," said Campbell. "They would go nuts. I want you to go home and write that story. About face! March!"

I went home, trembling with fear, sat down at my typewriter and tapped out "Nightfall." It appeared in the September 1941 *Astounding* and got the cover. It was my first smash hit, after nearly three years of trying. Robert Heinlein made it with his first story; A. E. van Vogt made it with his first story; Arthur C. Clarke made it with his first story. And I was almost as good: I made it with my sixteenth story.

"Nightfall" marked a turning point. From the moment of that sale, I never failed to sell a single word of fiction I wrote (though on rare occasions it took two or three submissions to do so).

Sometimes I wake up in the middle of the night screaming because I have dreamed that Theodore Sturgeon or Lester del Rey had walked into Campbell's office half an hour ahead of me that day. If they had, bang would have gone my Grand Mastership.

## 5: A Friend Insists

One of the stories I didn't sell—at first—was a novella entitled *Grow Old Along With Me.* I wrote it for *Startling Stories* at their request and, in the end, they rejected it. That happened in 1947, six years after "Nightfall." It rattled me. I decided that I had passed my peak—after all, I was twenty-seven—and I was sliding down the abyss to join Ed Earl Repp and Harl Vincent (two science-fiction writing idols of the early thirties).

Two years later, Doubleday decided to start a hardcover line of science-fiction. For that, they needed novels. I, of course, with my usual ability to keep my finger on the publishing pulse, knew nothing about it.

But I had a friend—Fred Pohl.

He came to me and said, "Doubleday is looking for a novel. How about the one you wrote for *Startling?*"

I said, "Fred, it's only 40,000 words. And it's a stinker."

He said, "So if they like it, you can lengthen it. And if you don't tell them it's a stinker, they might not find out."

But I didn't want to go through another rejection on the story, so I said, "I'd rather not submit it."

"I insist," said Fred.

I was not proof against Fred's quiet pertinacity, and I let him have the story. He let Doubleday have the story. Doubleday asked me to extend it to 70,000 words and took it. It appeared in January 1950 as *Pebble in the Sky,* and it has earned me money in each of the seventy-eight Doubleday statements I have received since then.

What's more, it got Doubleday into the pleasant habit of accepting my manuscripts as a matter of course. As of today, they have published one hundred and seven of my books and have several in press.

I'm sure that I would have been a reasonably successful writer on magazine short stories alone, but I would have been far poorer than I am today if I had not written my novels, and I would be nowhere near as well-known. In fact, if Fred had not insisted on submitting the story on that day in 1949, I doubt that I could ever have qualified for Grand Mastercraft.

## 6: A Critic Asks a Question

In 1957 I published my novel *The Naked Sun* in book form. It was a science-fiction mystery. Also, it had a rather understated love story in it, with a rather touching final scene between the lovers.

Damon Knight reviewed the book, and he wasn't in the least impressed by the science fiction or the mystery. (I suppose he's entitled to his opinion, but I don't suppose it very hard.) However, he liked the love story. "If you can write like that Asimov," he asked rhetorically in the course of his review, "why do you bother writing science fiction?"

To which I answered in a letter that appeared in the magazine in which the review had earlier appeared, "Because I love science fiction. No matter what else happens, I will never stop writing science fiction."

And then, in 1958, the very next year, I suddenly grew tired of science fiction. A sequel to *The Naked Sun* died in the typewriter, and I realized that I was anxious to write nonfiction. And yet how could I stop? I remembered my answer to Damon's question, and I simply couldn't go back on my profession of love.

It was while I hesitated that Robert P. Mills, then editor of *The Maga-*

*zine of Fantasy and Science Fiction,* asked me to write a monthly science column. I leaped at it, for that would enable me to write nonfiction and yet stay in the science-fiction field. It was the perfect Talmudic solution. The first science column appeared in the November 1958 issue of *F & SF,* and the column still continues to this moment, thirty-one years later. For twenty years, after that first column, I wrote mostly nonfiction. Mind you, I didn't give up my science fiction altogether. In that interval I wrote two novels and dozens of short stories, but compared to my earlier production it seemed like very little.

Had it not been for the *F & SF* column, which would not have come about, perhaps, without Damon's question and my answer, I would surely have been forgotten by the fans, thought of as another David H. Keller. That column kept me going till the founding of *Isaac Asimov's Science Fiction Magazine* in 1977 and Doubleday's insistence, in 1981, on my return to novels put me back in the mainstream. That column, by keeping me constantly in the public eye during the dry period, made it possible for me to earn Grand Masterness.

### 7: I Survive

Naturally, I had my vicissitudes. There was a hemithyroidectomy in 1972, and a mild heart attack in 1977. I survived both handily. And then, in the fall of 1983, my angina suddenly got so bad I could scarcely walk the length of the hall in my apartment-house.

On December 14, 1983, I had a triple bypass and came out of it in fine shape, thanks to a very clever surgeon. The morning after, I said to him, "The nurses tell me the operation went very well." And he answered, "What do you mean 'very well'? It was *perfect!*"

And so it seems to have been and had it not been so, there is scarcely any chance that an award would have been handed to a corpse. It is to that surgeon then that I owe my Grand Masterity.

* * *

The conclusion? Simple. I had nothing to do with it. If my parents hadn't brought me here; if my mother hadn't decided to let my name be; if a subway line hadn't existed at my very feet; if I hadn't wandered into Campbell's office at the right moment; if Fred Pohl hadn't insisted; if Damon Knight hadn't asked a question; and if a surgeon hadn't had a good day; I'd be left with nothing. As it is, I'm Grand Master—and I love it just as though I'd done it all myself.

# Acknowledgments

"Our Future in Education," copyright © 1988, by Nightfall, Inc.
"Filling the Brain Gap," copyright © 1988, by Northwestern Mutual Life.
"The Global Computerized Library," copyright © 1980, by Hearst Magazines, Inc.
"What Computers Won't Do," copyright © 1986, by Computer-World.
"The Future of Handicraft," copyright © 1988, by Nightfall, Inc.
"The Future of Chemical Engineering," copyright © 1988, by American Institute of Chemical Engineering.
"Men and Marriage," copyright © 1987, by the Conde Nast Publications, Inc.
"The Lure of Exploration," copyright © 1986, by the Boy Scouts of America.
"Our Second World," copyright © 1988, by Nightfall, Inc.
"All Aboard for Phobos," copyright © 1988, by Endless Vacation Publications, Inc.
"What Do We Do Next in Space?," copyright © 1986, by Family Media Publications.
"Adventure in Space," copyright © 1987, by the *New York Times.*
"The Distant Flights," copyright © 1988, by Final Frontier.
"The Telephone in Space," copyright © 1988, by Nynex Corporation.
"The Average Person as Astronaut," copyright © 1979, by the American Adventurers Association, Inc.
"Other Intelligent Life?," copyright © 1978, by Montcalm Publishing Corporation.
"Giant Jupiter," copyright © 1987, by R.C.S. Rizzoli Periodici, S.P.A.
"Pluto, the Constant Surprise," copyright © 1988, by R.C.S. Rizzoli Periodici, S.P.A.
"A Hole in the Sky," copyright © 1988, by American Airlines, Inc.
"Our Changing Perception of the Universe," copyright © 1984, by Nightfall, Inc.
"What is the Universe?," copyright © 1987, by Nightfall, Inc.
"The One-Man Revolution," copyright © 1979, by the Hearst Corporation.
"The Fifth Force," copyright © 1987, by American Airlines, Inc.
"Two at a Time," copyright © 1988, by American Airlines, Inc.
"Ozone," copyright © 1987, by Triangle Publications, Inc.
"The Ravages of Nature," copyright © 1987, by The American Legion.
"The Double Discovery of Evolution," copyright © 1980, by Triangle Publications, Inc.
"Master Lizard, the King," copyright © 1986, by Nightfall, Inc.
"The Hot-Blooded Giants," copyright © 1988, by American Airlines, Inc.
"The Absent-Minded Professor," copyright © 1980, by American Chemical Society.
"Playing It Safe," copyright © 1980, by American Chemical Society.
"The First Scientist," copyright © 1981, by American Chemical Society.
"Tough Luck," copyright © 1982, by American Chemical Society.
"To See Is Not Enough," copyright © 1980, by American Chemical Society.
"The Race for Honor," copyright © 1980, by American Chemical Society.

"Wishing Will Make it So," copyright © 1986, by Nightfall, Inc.
"Wizards," copyright © 1983, by Nightfall, Inc.
"Witches," copyright © 1984, by Nightfall, Inc.
"Curses!," copyright © 1985, by Nightfall, Inc.
"The Forces of Evil," copyright © 1983, by Nightfall, Inc.
"Monsters," copyright © 1988, by Nightfall, Inc.
"The Power of Evil," copyright © 1987, by Nightfall, Inc.
"The Devil," copyright © 1987, by Nightfall, Inc.
"Science Fiction Finds its Voice," copyright © 1988, by Nightfall, Inc.
"The Five Greats of Science Fiction," copyright © 1985, by Newsday.
"The Success of Science Fiction," copyright © 1988, by Nightfall, Inc.
"Science Fiction Today," copyright © 1988, by The Writer, Inc.
"The Feminization of Science Fiction," copyright © 1982, by the Conde Nast Publications, Inc.
"Back Through Time," copyright © 1986, by New York Times Corp.
"Our Shangri-La," copyright © 1987, by the Washington Post Corp.
"The Tyrannosaurus Prescription," copyright © 1988, by the Washington Post Corp.
"Ellis Island and I," copyright © 1988, by Nightfall, Inc.
"Seven Steps to Grand Master," copyright © 1988, by the Science Fiction Writers of America.